Sperm Wars

Dr Robin Baker was Reader in Zoology in the School of Biological Sciences at the University of Manchester from 1980–1996. Since leaving academic life in 1996, he has concentrated on his career in writing, lecturing and broadcasting. He has published over one hundred scientific papers and many books. These include *Baby Wars* and *Sex in the Future*. His work and ideas on the evolution of human behaviour have been featured in many television and radio programmes around the world. He has five children and has lived in Manchester since 1974.

Also by Robin Baker

Sex in the Future

ROBIN BAKER

Sperm Wars

Infidelity, Sexual Conflict, and
Other Bedroom Battles

BASIC
BOOKS

A Member of the Perseus Books Group
New York

Published by Basic Books,

A Member of the Perseus Books Group

Books published by Basic Books are available at special discounts for bulk
purchases in the United States by corporations, institutions, and other
organizations. For more information, please contact the Special Markets
Department at the Perseus Books Group, 2300 Chestnut Street, Suite 200,
Philadelphia, PA 19103, or call (800) 810-4145 x5000, or
e-mail special.markets@perseusbooks.com.

Library of Congress Cataloging-in-Publication Data is available.

ISBN-13: 978-1-56025-848-3

ISBN-10: 1-56025-848-9

Book design by Intype London Ltd

20   19   18   17   16   15   14   13

To

K
E
J
F, Y
A, J, K, P, P, R, S, 'W'

and

Thomas, Howard, David, Nathanial, Amelia
and the baby who was never named

Contents

Preface   xi
Introduction to the 2006 Edition   xix

1.   The Generation Game

1: Great Uncle Who?   1

2. Routine Sex

2: Normal Service   6

3: The Wet Sheet   16

4: Topping Up   27

5: Conception   33

3. Sperm Wars

6: A Chance Affair   38

7: A Sperm War   51

4. Counting the Cost

8: Doesn't He Look Like His Father?   58

9: Making Mistakes   65

10: Licking Infidelity   73

11: Checkmate   80

5. Secret Anticipation

12: A Double Life   90

13: Multiply, But Don't Divide   97

14: A Wet Dream   105

6. Successful Failure

15: Home for the Day   109

16: The Stress of It All   115

17: How Forgetful   129

7. Shopping Around for Genes

18: Spoilt for Choice   137

19: Fair Exchange   153

20: Tasteful Display   161

21: An Abandoned Selection?   171

8. The Climax of Influence

22: Finger on the Button   191

23: Dark Secret   200

24: Another Successful Failure   205

25: Correcting Mistakes   212

26: Putting It All Together   222

9. Learning the Gropes

27: Practice Makes Quite Good   236

28: Rough and Tumble   245

29: How to Con   264

10: One Way or Another

30: Best of Both Worlds   276

31: The Coming of Women   294

32: The Tenth Tonight   307

33: The Predator   319

34: Soldier, Soldier   330

35: Men Are All the Same   340

36: Exquisite Confusion   349

11. Final Score

37: Total Success   366

Preface

Sex and reproduction occupy a major part of people's time – not so much the doing, more the thinking and talking. Despite all of this attention, most people find their own sexual activities, responses and emotions the most baffling of all aspects of their lives. Consider the following.

Why, in the midst of a perfectly happy and satisfying relationship, do we sometimes get an incredibly strong urge to be unfaithful? Why do men inseminate enough sperm at each intercourse to fertilise the entire population of the United States – twice over? And why, then, do half of them dribble back down the woman's leg? Why should we feel like sex so often when most of the time we don't want children? Why, when we least want children, do our bodies apparently let us down and produce them? Why, when we do want children, do our bodies apparently let us down and *not* produce them? Why is it so difficult to know the best time to have sex in order to conceive – or not to conceive? Why is the penis the shape it is and why do we thrust during intercourse? Why do we get such strong urges to masturbate, and why do some of us have orgasms while asleep at night? Why is the female orgasm so unpredictable and so difficult to induce? Why are some people more interested in sex with members of their own sex?

These are just some of the questions to which most people, if they are honest, have no sensible or at least consistent answer. None the less, riding on the wave of a revolution in sexual understanding which began in the 1970s but which did not really gather momentum until the 1990s, these are questions which this book sets out to answer.

So far, this revolution in the interpretation of sexual behaviour has been the sole prerogative of academics – evolutionary biologists, to be precise. In this book, my aim is to bring the new interpretation for the first time within the reach of a much wider audience.

The potential exists to revolutionise the way we all think about sex. My ambition is to help the revolution on its way. The main message from this revolution is that our sexual behaviour has been programmed and shaped by evolutionary forces which acted on our ancestors – and which still act on us, even today. *The main thrust of these forces was directed at our bodies, not our consciousness.* Our bodies simply use our brains to manipulate us into behaving in a way dictated by our programming.

The central force that directed this programming was the risk of *sperm warfare*. Whenever a woman's body contains sperm from two (or more) different men at the same time, the sperm from those men compete for the 'prize' of fertilising her egg. The way in which these sperm compete is akin to warfare. Very few (less than 1 per cent) of the sperm in a human ejaculate are the elite, fertile 'egg-getters'. The remainder are infertile 'kamikaze' sperm whose function has nothing to do with fertilisation as such but everything to do with preventing sperm from another man fertilising the egg.

Sperm warfare is a story in itself, but it also has wide-ranging consequences at all levels of human sexual behaviour. In part consciously, but much more importantly subconsciously, all of our sexual attitudes, emotions, responses and behaviour revolve around sperm warfare, and all human sexual behaviour can be reinterpreted from this new perspective. Thus, most male behaviour is an attempt either to prevent a woman from exposing his sperm to warfare or, if he fails in this, to give his sperm the best chance of winning that warfare. Most female behaviour is an attempt either to outmanoeuvre her partner and other males, or to influence which male's sperm have the best chance of succeeding in any war that she promotes.

For each one of us, there was a critical moment, some time in the past, when one of our father's sperm entered one of our mother's eggs and we were conceived. That event unleashed a complex set of instructions for development. Those instructions were inherited half from our father and half from our mother and eventually produced the person we are today. If our father and mother hadn't had sex when they did, with whom they did, having prepared for it as they did, we would never have existed.

Behind every conception lies a story. But the details of these stories are rarely known. How many of us know, for example, whether our mother climaxed at our conception and, if so, whether she did so before, after or at the same time as our father? Did either our father or mother masturbate in the days or hours preceding our conception? Was either of them bisexual or was either of them being unfaithful to their partner at the time? When we were conceived, did our mother contain sperm from only one man or did she contain sperm from two or even more? Is the man we call our father actually the man who produced the sperm which fertilised the egg from which we developed?

These things *will* have made a difference to our personal origins, and an understanding of the precise ways in which they did is one of the most interesting outcomes of the new revolution.

Most people, of course, are conceived during an act of routine sex between a man and a woman who are living together in some form of long-term relationship. This is true now and has probably been true for at least the last three or four million years. Such conceptions might seem humdrum, but even routine sex has its surprises, as I hope this book will illustrate. And for those one in five or so people who are *not* the product of routine sex, there is an even more interesting story behind their conception. Many such stories are told in this book.

In 1995, Dr Mark Bellis and I published a book called *Human Sperm Competition: Copulation, Masturbation and Infidelity*. In that book, published by Chapman & Hall, we presented the results of recent biological research, much of it our own, about the repercussions for human sexuality of the risk of sperm warfare. We argued that almost every aspect of human sexuality, including much of the familiar and humdrum, owed its characteristics to the occurrence, or at least the risk, of sperm warfare. If you wish to read a scientific justification for the ideas and claims made in *this* book, then read *Human Sperm Competition*. Of necessity, that book is full of jargon, data, graphs and tables, which inevitably distance it from most people's experience. Nevertheless, it contains an interpretation and explanation for all the sexual behaviour that is very much within most people's experience – behaviour that often seems irrational and inexplicable. The research shows, though, that sexual behaviour, in all its mundane, embarrassing, pleasurable, risky, criminal, amoral and exotic forms, does obey fundamental rules.

In order to illustrate these rules and to bring the behaviour to life, I have written this book as a series of fictional scenes. Every scene involves some form of sexual conflict – between males, females or, most often, between males and females. Many scenes also involve one or other facet of the sperm warfare which I argue throughout the book is the underlying element in all sexual behaviour. Each piece of fiction is followed by an interpretation of the behaviour just witnessed, from the viewpoint of an evolutionary biologist.

The fictional scenes show people acting out those sexual strategies that have been the main targets for investigation and interpretation in recent years. While I have drawn evidence about how people behave from a wide range of scientific surveys and experimental studies in which, in total, many thousands of people around the world have taken part, the stories themselves, naturally enough, are contrived. Their aim is, after all, to show people experiencing particular costs and benefits of their sexual behaviour so as to neatly illustrate the subsequent evidence and interpretation. My challenge was to create characters and scenarios that offered a reasonably entertaining and plausible piece of fiction, while at the same time reflecting that evidence.

In writing these stories, I have drawn on less global information than do the studies and experiments themselves – and although some stories originate in news reports in the media, most derive from events in my own life and the lives of close friends and family. All of the scenes have been inspired by actual events. Even so, my friends should not waste their time trying to guess identities, nor should the reader waste his or her time trying to identify particular news stories. Every character is an amalgam, and every story a mosaic of several different events. Moreover, every character described could be of any race (and almost any nationality), and every scene could be set in almost any country.

Not every event in every scene is interpreted on the spot, but every element of behaviour mentioned is interpreted in its own right somewhere in the book. For example, I have dedicated two scenes to masturbation, one to male (Scene 12) and one to female (Scene 22). After each of these scenes, I discuss the function of masturbation. Elsewhere, scenes of sexual behaviour often involve characters masturbating or having masturbated, without my interpreting their behaviour then and there. But once the functions of masturbation have been gleaned from their dedicated scenes, the involvement of masturbation in other scenes should be clear.

In writing my interpretations, I have tried not to lapse into the academic style that is my natural bent. I have tried to avoid too much mention of numbers and, where a particular explanation is complex, I have tried to give a brief and entertaining version even if it means sacrificing academic precision. I have also avoided using the words 'probably' and 'possibly' on many occasions when, strictly speaking, they were needed. Any reader with a scientific background who becomes frustrated by the lack of academic rigour in my text should seek the information and explanation they require in the treatise I wrote with Mark Bellis.

Not all of my fellow academics will agree with my interpretation or even with the details of what is going on between men and women, between sperm and the female tract, between sperm and the egg, or between sperm themselves. There are people, eminent in their fields, who will consider this whole book to be a fiction, interpretations as well as scenes. So be it. The point is that I have opted to tell a story, a story based on a genuine academic interpretation of recent research. Apart from this self-imposed starting point, my primary consideration thereafter is that the story should be consistent and interesting. I have not even begun to present all sides of other people's views. To have done so would have made the book confusing, over-long, even dull. Other people's interpretations are discussed and appraised at great length in *Human Sperm Competition*, in which Mark Bellis and I say precisely why we think the story I present here is the best currently available. Having argued the case there, I feel free here to tell the story as simply and as entertainingly as I can manage.

One of the problems I encountered in writing this book is that much of the behaviour I am attempting to interpret requires a totally explicit picture of what happens. Many of the scenes and details I describe could, in another context, be considered pornographic. I have tried not to be gratuitously explicit and hope that, for every scene or detail a reader might find embarrassing or titillating, the explanations that follow will provide adequate justification.

I have a further problem. A large part of the behaviour I describe and interpret is to many people at best amoral and at worst criminal. In my view it is most important that I do not strike any moral stances. As an evolutionary biologist, my aim is to interpret human behaviour without prejudice or criticism. The danger is that many people will interpret my lack of criticism of certain forms of behaviour as meaning that I condone or encourage that behaviour. However, as I explain in Scene 33 in relation to rape, the first step in dealing with antisocial behaviour is to understand it – and that and nothing else is the aim of all my interpretations.

This book could never have been written had it not been for my scientific collaboration with Mark Bellis. I owe him a great debt. For seven years, from 1987 until 1994, we discussed, investigated and argued about every aspect of human sexuality. We did not agree on everything, but to a surprising extent one or other of us eventually became persuaded by the other's arguments – sufficiently persuaded, that is, to collaborate in writing the academic book that is the scientific foundation for this one. Although Mark will not agree with all of the ideas I present here, some of which have emerged since he left Manchester University to start his important work on the epidemiology of AIDS and other sexually transmitted diseases, most of the ideas presented here are as much his as mine. He should not be held responsible, however, for the way any of these ideas are presented. Naturally, he is also totally free of blame for my fictional representations.

I am also grateful to Fourth Estate, particularly Michael Mason and Christopher Potter, for showing what I consider to be great courage in publishing this book – not because of the subject-matter but for their confidence that I could actually deliver a saleable product. Before they offered their support, I had no grounds for thinking I could write a book such as this, previously having written only academic texts. I hope the end result has justified their initial confidence.

By far my biggest thanks must go to my partner, Elizabeth Oram, who has encouraged and helped me at every step of the way. She did so despite considerable physical discomfort – the conception, gestation and delivery of this book coinciding with the conception, gestation and delivery of our second child.

The book was conceived over breakfast one Saturday in October, three months to the day after the conception of our child. Without her immediate encouragement, I would never have dared to begin this project.

As Liz and the book grew, she applied her considerable writing and editorial skills to prevent me from making many of the mistakes that would have exposed me as the novice I am in the popular genre. First, she saved me from my urge to be even more graphic in my description of sexual scenes than I have. If the final results are interesting and explicit – but still tasteful – they owe much to her restraint and advice. Secondly, she saved me as much as she possibly could from being outlandishly pompous and didactic in writing the interpretations. If the book still contains elements of pornography or pomposity, the fault lies not with her advice but with my stubbornness and reluctance to kill words that I conceived with such difficulty.

Finally, as the race to deliver the manuscript before Liz delivered our baby gathered momentum, she did everything she could to make sure I didn't sacrifice quality for speed. In order to free time for me to finish the manuscript she allowed me to shirk many of my paternal and household responsibilities. Not once was I made to feel guilty. Moreover, she read and reread successive drafts of each scene and interpretation whenever I asked, even late at night. Throughout all the pressure, when I could so easily have decided that this or that sloppy paragraph 'would do', she never once allowed her editorial standards to drop.

Her final effort was superhuman! When I failed to complete the manuscript by the expected delivery date of our baby, she hung on and refused to give birth for a further ten days. Thanks to the time she freed for me, I managed to deliver the manuscript to Fourth Estate just before our first daughter, Amelia, was born.

Introduction to the 2006 Edition

The first twenty-six years of my working life were spent as a university scientist, giving lectures, carrying out research, and writing a series of academic books on a variety of subjects. As for most academics in my position, a successful publication was one that sold—not really to people—but to a reasonable proportion of university libraries in the English-speaking world. There they would sit, gathering dust, waiting for somebody to need to consult this or that piece of information or viewpoint. Such books aren't usually read as such. They are picked at—often nitpicked at—by other academics, from professors to students, as often as not looking for flaws rather than facts. Then, ten years ago now and a trifle disgruntled at having so much effort reach so few people, I turned my hand to writing popular science. The hope was that the words might actually be read outside of ivory towers—and maybe even by people who spoke languages other than English. To my great pleasure, both hopes were realized and never again could I contemplate reverting to writing for academia.

Sperm Wars had a subject matter—sex, fertility, and reproduction—of interest to all and was meant to both entertain and inform; to help people think about their own and others' actions; to give a new perspective on the sexual behavior and urges that everybody experiences at some time in their lives. Not only those urges that all societies consider normal and acceptable but also those that many don't. What I never really contemplated while writing was that some people might find the book reassuring. Ironically, this seems to have been one of its most common roles.

I had expected mail: from incensed academics accusing me of prostituting my scientific research; from zealots condemning me for seeming to condone via objective explanation behaviors that are or are considered to be antisocial, illegal, or even sinful; and from male and female chauvinists angry because I appear to justify what they see as the heinous actions of the opponent sex. None of this really happened. Academics preferred to vent their spleen in the introductions to their own books rather than confront me directly—and zealots and chauvinists by and large either appreciated what I was saying or saw no cause to react. Instead, my mailbox was, and still is, filled with letters from men and particularly women from all over the world thanking me for explaining why they had at some time or other behaved in what, until reading my book, they'd considered to be inexplicable ways. Most of these letters concerned infidelity or promiscuity; others were about masturbation or nocturnal climaxing. The response wasn't what I'd anticipated, but much appreciated nonetheless.

The letters came from all over the world because, thanks to the efforts of my energetic and accomplished agent, Laura Susijn, foreign publication rights to *Sperm Wars* were eventually sold to twenty-one different publishers, from Canada and Russia in the North to Australia in the South, from the United States in the West to mainland China and Japan in the East. Altogether, it has so far been published in a gratifying twenty different languages.

Such worldwide translation and publication has allowed a fascinating insight into the reaction of different nationalities to the open, nonjudgmental discussion of human sexual behavior found in the book's pages. In the UK it quickly made the Sunday *Times* bestseller list, peaked at number 7, and stayed for a few weeks. There was a similarly enthusiastic response to the German, Polish, and first Chinese translations. In contrast, the Danish and Swedish translations seemed to arouse much less excitement. Maybe frank and liberal-minded discussion of sex was nothing new in Scandinavia by 1996! The response in the United States, however, was quite different from anywhere else and took me completely by surprise.

The weeks after *Sperm Wars* was published in America I performed what seemed an endless state-to-state round of radio interviews. Those in California and New York followed the pattern to which I'd become accustomed in Europe—lively, open, inquisitive, and only occasionally bawdy badinage about infidelity and other intimate details of people's sex lives. Lulled into a state of false confidence, I moved on to other states—which I prefer not to name.

The first and most bizarre shock came from a radio station that enthusiastically offered a fifteen-minute conversation about *Sperm Wars* with its celebrity host—with the proviso that neither I nor he used the word "sperm," which was considered offensive. After that, although never quite so strange, station after station placed this or that limitation on what I could say. Describing the book's content and message was more often than not well near impossible. Worst of all (for me) was the station that, I assumed to be well aware of the nature of the book, offered an hour-long interview. Five minutes in, the host gave an on-air plea of "help" to the technicians around him. Two minutes later he pronounced that he just couldn't handle the conversation and cut me off, replacing my interview with fifty minutes of canned music. The reason? It wasn't because I'd sworn or been coarse, or that I'd talked about genitals or orgasms. It wasn't even because I'd used the word "sperm." It was because I had listed, soberly and earnestly, the reasons why a woman sometimes finds herself having sex with more than one man in the space of a few days, the behavior that generates the sperm wars that were the basis of my book.

Sperm Wars sold well enough in the United States first time round. However, whether due to the prudishness, embarrassment, and reluctance to openly discuss sexual matters that I encountered on my radio tour in 1996 or not, it reached far fewer people than in parts of Europe and Asia. Certainly not as many as I and its original publishers had expected. So why relaunch it in America now, nearly ten years after its initial publication?

It isn't because human reproduction is beginning to change. Until the end of the twentieth century, our basic sexual behavior had stayed more or less the same for thousands, maybe over a million, years. However, as I described in another book, *Sex in the Future,* the greater availability and reliability of reproductive technology will one day influence much of the behavior detailed in *Sperm Wars.* DNA paternity testing, in particular, is beginning to reduce the chances of a man being tricked into unknowingly raising another man's child and hence impacting on the costs and benefits of infidelity. For the moment, though, this impact is still minimal. Nor is this relaunch of *Sperm Wars* really because the book has been accredited "classic" status as has been the case elsewhere in the world—in Britain, France, Poland, Czech Republic, and China to name a few—where relaunches have already taken place or are in process. It isn't even because for a time in the States the original became a collector's item. Not long after the first version had disappeared from the nation's bookstores, second-hand copies were being offered for sale on one of America's biggest Internet bookshops for ninety dollars a copy, having originally retailed at around twenty dollars. Instead, the main reason is the feeling that the average American in 2006 is more ready, able, and maybe even eager, to tackle the sorts of questions and explanations with which this book deals than he or she was in 1996. Ready—and less easily embarrassed.

Over the last few years there has been both a runaway success of national chain stores specializing in sex and a never-ending spate of news, articles, and features about Americans' obsession with "sex toys." The toys, themselves, aren't really new. In principle they are still no more than aids to solitary male or female masturbation or to couples for making their sex life more varied. What is new, though, is the fact that these toys are selling by the zillions to a broad segment of the population—not just to the inhabitants of New York and San Francisco! The current batch of American under-thirties has even been branded "Generation Porn." Sex is everywhere, from Web to television, and it would be encouraging to think that it reflects or at least precedes the development of the healthy, informed, and nonjudgmental attitude that *Sperm Wars* is intended to promote.

There is, of course, still a way to go. Even the most optimistic American liberal would admit that the United States is still less relaxed over sexual matters than parts of Europe and Asia. So, too, would the less fortunate of its public figures. Many politicians, from ex-presidents down, might wish that their electorate were more similar to those in parts of Europe where having a lover, the more attractive and famous the better, is an almost mandatory part of a successful image. They might wish that "doing what comes naturally"—the subject matter of *Sperm Wars*—was better understood by the average American.

I wait with interest to see if the *Sperm Wars* message engages the current generation in the United States even more than it did the previous. In the meantime I thank John Oakes and Avalon Publishing Group both for their faith in my book and their faith in its potential American audience. I also thank them for helping me meet a personal yardstick, one that emanated from the professor who supervised my doctoral studies many years ago. A wise and deliciously eccentric man, Howard Hinton had a smattering of Mexican genes and a rich store of memorable maxims. Which of these he'd conceived himself and which he was simply passing on I was never quite sure, but two in particular have become relevant to my *Sperm Wars* experience.

His first maxim was that academic opponents never change their mind, so the only victory one can ever hope for is to outlive them. The second was that if anything one writes isn't still being talked about ten years later then it wasn't worth writing in the first place. Whether I shall outlive those few academics incensed by *Sperm Wars* is for the future to reveal—but the relaunch of the book in the United States ten years after it was first published is now a matter of record. Whether this edition has even more meaning to the American book-reading public the second time around remains to be seen, but I at least have the pleasure of knowing that my now-deceased mentor would have given it his irreverent blessing.

Robin Baker
August 2005

1

The Generation Game

Great Uncle Who?

The faces in the creased brown photograph stared impassively at the woman, their gazes spanning the hundred years between them. She loved this photograph and often asked to see it when she visited her grandmother. The faces belonged to three young children, all long dead, frozen in time by some ancient camera at a moment early in their lives. They were standing in a line, tallest and oldest on the left, shortest and youngest on the right. The two boys at either end were aged about ten and two, the pretty girl in the middle about five.

Whenever the young woman looked at these faces, she sensed a continuity with the past that she never experienced at any other time. The photograph showed her great-grandmother with her two brothers. But, with very little stretching of the imagination, it could be her looking out from the photograph, not her great-grandmother. The resemblance between them as children was uncanny. Her grandmother called it 'the family face', so many of their clan having the same bone structure and eyes.

The woman looked at the photograph a little longer, then asked her grandmother to tell her the story of their family 'just one more time'. Before speaking, the old woman fumbled to the front of the album and took out a large sheet of paper. This family tree was her pride and joy and she loved showing it and the photographs to her many grandchildren.

The young woman concentrated hard as her grandmother spoke, determined this time to remember what was said. She knew that one of the boys in the photograph had not lived long enough to have children. Her great-grandmother, however, had not only survived but had also escaped the poverty of her family background. She had been a pretty child who had grown into a beautiful young woman, chased by all the young men in the village. One day, while working as a servant in a large household, she had fallen pregnant to the owner's son. The baby was her grandmother, the teller of the story.

Instead of being disowned and sent away, her great-grand-mother was welcomed into the family. Everything happened so quickly that, despite gossip, nobody ever knew for sure that the baby had been conceived illegitimately. The young couple then lived together in relative comfort for the rest of their lives and produced four more children. All were boys and, unusual for their generation, all had survived.

The grandmother then pointed to the oldest boy in the photo-graph, her uncle. He had not been as lucky as his sister. Failing to escape the poverty into which he had been born, he had worked hard all his life. Like his sister, he also had five children. Three had died as babies, and one of the survivors, a boy, had been killed in the war when only eighteen years old. The other survivor, a girl, was infertile and died alone, in her fifties, a few years after her partner. The youngest boy, the one with the bright eyes and smile, had died of measles about two years after the photograph was taken.

The young woman pored over the family tree with her grand-mother. The tree had the shape of a pyramid: three names at the top, the three young children in the photograph, and about fifty at the bottom, the woman's own generation. Then suddenly she noticed something that had never occurred to her before: every single one of the fifty people in her generation traced back to her great-grandmother, the pretty girl in the picture. Not one, of course, traced back to either of those two boys.

The young woman bent forward to look at the family tree more closely. She was looking for others who, like the two boys, had no living descendants and whose lines on the tree therefore ended in mid-air. The most conspicuous was one of her grandmother's brothers, the great-uncle whose name she could never remember but who was reputed to have had a very strangely shaped nose. She spotted two more lines ending in mid-air before her stance became too uncomfortable. Unable to bend forward any longer, she straightened up and turned away from the paper and photographs. As she did so, the baby in her womb kicked. She winced, then smiled and held her stomach. At least *her* line wasn't going to end in mid-air.

~

Our personal characteristics depend on our genes – chemical instructions as to how we should develop and function. These instructions are packaged in sperm and eggs and passed down our family tree, finally reaching us via our genetic parents. And we inherit more than our 'family face' via these genes: we also inherit many aspects of our physiology, psychology and behaviour, including much of our sexual behaviour.

This book's task is to work out why we behave sexually as we do. Our approach is simple. We shall ask why it is that people with some sexual strategies (patterns of sexual behaviour) are *more successful reproductively* than people with others. Our measure of success will be the number of descendants people achieve – because this is what shapes future generations.

Families and populations become dominated by the descendants of their most successful ancestors. They also become dominated by those people's characteristics. In the scene we just saw, the younger woman's generation was dominated by her great-grandmother's face, not by great-uncle who's nose. For all she knew, her generation was also dominated by a 'family sexuality', passed on to so many people by the dynasty founders, her great-grandparents. Nobody will have inherited directly the sexuality of great-uncle who. Whatever his sexual strategy might have been, it was unsuccessful and he left no descendants to inherit that strategy.

It is irrelevant to our generation whether people in the past wanted many children and grandchildren or whether it just happened. The only factor to shape our characteristics is who in the past had children (and how many) and who did not. The great-grandmother and great-grandfather in Scene 1 were probably most dismayed when their sexual fun produced a child. But if it hadn't, the younger woman and her fifty or so family contemporaries would not have lived. In effect, each generation plays a game in which its members compete to pass their genes on to the next generation. Each generation has its winners, like the pretty girl in the picture, and each has its losers, like her two brothers and great-uncle. We are the descendants of the winners, the people whose sexual strategy paid off.

The generation game has not ended. It will continue for as long as some people in a generation have more children than others. In our own generation, the game is as active and cruel as ever. It will still be the genes of those among us who produce most descendants that will characterise future generations, not the genes of those who produce few or none.

Whether we know it or not, whether we want to or not, and whether we care or not, we are all programmed to try to win our generation's game of reproduction – we are all programmed to pursue reproductive success. Our successful ancestors have saddled us inescapably with genetic instructions which tell us not only that we *must* compete, but also *how to*. Inevitably, some of us will have had more successful ancestors than others, so that even in our generation there will be some people who have inherited instructions for potentially better strategies. When our generation comes to work out its final score, some people will have done better than others. We are about to start investigating why it is that some people are more successful than others in life's generation game.

2

Routine Sex

SCENE 2

Normal Service

It is late on a Saturday night and a man and woman in their late twenties are getting ready for bed. As they drift around their rooms, attending to the minutiae of life, they are naked. For them, this is usual and of no sexual importance. They are no longer excited by simply being naked in each other's presence. In fact, they now scarcely notice each other's bodies. As it is Saturday night, they know they will have sex before they go to sleep. Yet, as they vacantly pursue their separate routines, there is no hint of foreplay, even when on occasion their paths cause their bodies to brush past each other.

It is a week since they had sex – last Saturday, in fact. Four years ago, when they first met, they had sex at least once a day (except during her menstrual periods, when neither of them was particularly keen). In those early days they would have ridiculed the possibility of intercourse only once in a whole week. Now, once a week had become more and more common, even though their usual routine was still to have sex twice a week. Until, that is, two months ago when they had given up using contraception.

Not that they were in any rush to have children. They hadn't yet contemplated the earnest nightly conception campaigns that some of their thirty-something friends had delighted in describing to them. Rather, they preferred to leave it to fate (and so far fate had decreed 'no conception'). They had both found mild sexual excitement in the possibility of conception and for a while their rate had returned to three or four times a week. This week, however, had been different. A couple of separate nights out and, if they were honest, an unexplained coolness between them had conspired against their ever quite getting round to sex. The usual warmth of their relationship had not fully returned until this Saturday morning as they drove on a pre-arranged visit to her sister. Even now, as they eventually got into bed, they could both still feel the legacy of the week's coolness. It was with some tentativeness that the man made his first faltering contacts with his partner's bare body. Once started, however, they quickly slipped into their usual routine.

He begins by gently kissing her face and stroking her breasts. Then they kiss deeply. He strokes her legs to her knees. After a while, he moves down and sucks her nipples. All this time, she cursorily strokes his back and buttocks. Tonight, as is often the case, she cannot concentrate and her mind keeps slipping back to conversations with her sister earlier in the day. She is jolted back to the present when he places his hand between her legs, moves her longest pubic hairs, opens her lips and inserts a finger to check if she is wet. He thinks she is ready. She knows she is not and winces at the prospect of unlubricated penetration. She moves her hand, finds his penis and gently squeezes, in part to see how ready he is but primarily to delay his moving into position. Briefly, her ploy works. He pauses to savour the sensation and responds with half-hearted massage of her genitals. Even though his massage misses her clitoris by a centimetre, he detects (or imagines) an increase in wetness on his finger inside her vagina. He moves his hand and begins to shift his body into the missionary position. She keeps her hand on his penis, and when the moment comes helps to guide its swollen tip into position. She leaves her hand between them for a few seconds to stop him pushing too hard, too soon (she is still nowhere near moist enough). Then, she has no alternative but to abandon the act to him. It takes a while before his gentle working backwards and forwards makes her lubricants really start to flow and his penis is able to enter fully.

Until she was lubricated, the woman had focused her mind on his and her genitals and the mechanics of penetration. But once she is lubricated and he begins the routine of thrusting, her mind drifts back to her sister. Her attention returns to the present only when he makes an uncomfortable movement. Despite her abstractedness, years of practice allow her to time the quiet noises in her throat to the man's thrusts. Then, suddenly, her mind jumps back to Wednesday night and the man who had flirted with her when she was out with a group of her female friends. Now, in her mind, it is him on top of her. Her heart speeds up, her breathing quickens, and her noises get louder. But just as her fantasy begins to take shape and she feels she might even come, her partner makes a particularly awkward thrust. Her fantasy disappears. The moment has gone, and the next second she realises he is ejaculating. She makes a sound for each of his contractions, then relaxes with him as his penis shrinks inside her. Impatient for him to remove his now dead weight, she coughs, gently. His limp appendage is ejected, he moves off her and they slip into their usual post-coital embrace. Both feel guilty at not having made more effort for their partner's sake and both feel depressed. Briefly, they exchange untruths over how pleasurable everything had been before eventually drifting into post-coital sleep.

~

For most people, the commonest situation in which they have sexual intercourse is at home and with a long-term partner. Rapidly, such intercourse becomes a matter of routine within the relationship; but, routine as it may be, it plays a surprisingly important role in the man's and the woman's pursuit of reproductive success.

This chapter consists of Scenes 2 to 5, each of which explores one or more facets of intercourse within a long-term relationship. Whereas most of the scenes in the book will involve different characters in different situations, these first few (Scenes 2 to 7) follow a single couple. We stay with them and follow their routine sex until the woman conceives and we have unravelled the full story behind her conception.

While interpreting these first scenes, I take the opportunity to explain some of the basics of human sexuality. Many of these will be well known to readers, but I guarantee a few surprises. Some of the descriptions may seem a little detailed, but this detail will help later when we discuss some of the more interesting aspects of human sexual behaviour – such as male and female masturbation and the female orgasm.

Anyone who has ever lived in a sexual relationship for a few years should find familiar elements in the scene we have just witnessed. In fact, it is so familiar that there is a danger of missing the subtlety of both people's behaviour. Here we meet a couple who have had penetrative sex perhaps five hundred times in the four years of their relationship. Yet not one of those inseminations has led to pregnancy. Of course, they have been using contraceptives, but from time to time they will have been careless and the woman could have become pregnant, but she didn't. Now they have stopped taking contraceptives, but she still hasn't conceived.

Clearly, they have not repeated this particular act five hundred times in order to have children. Nor are they unusual in this. The average man and woman, no matter whether they live in the bush of the Kalahari desert or in a multi-bedroomed executive house, will have penetrative sex about two to three thousand times in their lifetime. Yet, even without modern contraception, most people have fewer than seven children. This works out at about five hundred inseminations to produce each child, though the precise figure is unimportant. Whatever the fine details of the arithmetic, the conclusion is inescapable. From the viewpoint of reproductive success, people do not have routine sex primarily to produce children.

Nor are humans unique in this respect. In fact, compared to other primates, we are probably fairly average in terms of the number of inseminations per offspring. We pale into insignificance compared with pygmy chimps, who seem always to be having sex. Outside of primates, the lion, which takes three thousand inseminations to produce another lion, also beats us fairly easily. Some birds may mate only a handful of times for each nestling, but others are about the same as us and mate hundreds of times to produce each young bird. So why do we, and all these other animals, mate so often if not to procreate? How does routine sex help both men and women in their pursuit of reproductive success?

The explanation usually trotted out is that we (and presumably, therefore, all of these other animals) have sex because we enjoy it, because it brings pleasure. But is that really true? Look at the couple in our scene again. Of course, in those first few weeks of their relationship when they were having sex every day, the penetration, the contact and even the prospect of just being naked together was exciting to both of them. Also, of course, even since that first flush of excitement, one, the other or both of them will occasionally have gained real pleasure from an intercourse. But recently, our couple have experienced such pleasure less often. In the scene we have just witnessed, neither partner was particularly looking forward to sex and, if they had both been honest, neither of them actually gained that much pleasure from their intercourse.

The woman certainly didn't. The whole process was uncomfortable, even marginally painful, and almost totally unrewarding. She had experienced far more sexual excitement from simply flirting with another man the previous Wednesday than she did from full sex with her partner this Saturday. As for the man, he was bored during foreplay, irritated at having to insert himself into an unlubricated vagina, and both bored *and* irritated waiting for her to become aroused while he was thrusting. He had a brief pleasurable sensation in the seconds before ejaculation, but almost immediately afterwards sank into a guilt-ridden depression. Moreover, not only had the pair gained little pleasure from their union, they had both known they wouldn't even before they started.

So why did this couple have sex this particular Saturday night, and why will they do it again and again over the weeks, months and even years to come?

In its most general sense, the explanation for routine sex is tautological. Men's and women's bodies are programmed by their genes to seek sex with their partners at intervals, as a matter of routine, whether their brains can see a good reason for it or not. Why? Because routine sex can actually make a difference to the number and quality of children, grandchildren, great-grandchildren and so on that men and women might have. It can do so despite leading to conception on only one out of every five hundred or so occasions. Moreover, it does so without the conscious brain knowing and, most often, without it actually caring.

So what is the big benefit of routine sex within relationships that needs no involvement from the conscious brain? The precise answer depends on whether you are male or female, and introduces us to the first example of a theme that runs throughout this book: *what is best for one partner is very often not best for the other.* In this case, what men's bodies are trying to do is maintain a population of sperm inside their partner. What women's bodies are trying to do is confuse the man so he never knows, either consciously or subconsciously, the best time to inseminate her.

Some female primates, such as chimpanzees and baboons, actually advertise when they are most fertile each month by developing large and conspicuous red, crusty swellings around the anus and vulva and sometimes red skin on the chest as well. Male baboons and chimpanzees get turned on by these signals, and are much more interested in sex with the female when she is at this peak of beauty! The males compete for a female most strongly during the few days when she is most fertile, and each male does his very best to prevent her from mating with other males. Often he gives up doing other things, like feeding, just so that he can keep a close eye on her.

In contrast, many other primates, particularly those which, like humans, form monogamous relationships (gibbons, for example), hide rather than advertise their fertility. Why? Well, if the male doesn't know when the female is most fertile, he can't guard her so intensively. After all, he can't give up eating and sleeping indefinitely. So, by hiding her fertility, a female primate gives herself much more control over when and by whom she conceives. In particular, she makes it easier for herself to be unfaithful to her partner if she ever wants or needs to be. This works for women just as well as it does for gibbons.

The elegance and effectiveness of a woman's ability to hide her fertility from men is breath-taking. On the one hand, her body creates an environment in which conception is relatively easy but only if the timing is absolutely right. On the other hand, her body gives absolutely nothing away to the male that could help him to get that timing right. The details of this strategy for confusion are fascinating.

First, as a general rule, a woman's body allows sperm to remain fertile for no more than five days after being deposited inside her. Secondly, sperm seem to need about two days inside the female to reach peak fertility. Thirdly, women produce just one egg per menstrual cycle, but this egg dies within a day of being produced by her ovary. What all this means is that for a man to have any chance of fertilising a woman, he has to inseminate her at least once during the period from five days before she ovulates to about twelve hours afterwards. To have the best chance, which still isn't very high (about one in three), he must inseminate her about two days before she ovulates. A day or so either side of this optimum time and his chances decrease dramatically.

At first sight, it might seem that all a man has to do is note when his partner starts to menstruate, wait twelve days, and then inseminate her. That way, his sperm would reach peak fertility two days later on day 14 of her cycle. This is the day that many people assume marks the peak of a woman's fertility. However, the woman's body easily outsmarts such simple arithmetic: a predictable menstrual cycle is a rarity rather than the norm, and only occasionally does a woman ovulate on day 14. The key to her body's strategy is variability, and hence unpredictability.

The total length of the menstrual cycle, from the beginning of one period to the beginning of the next, can be anything from about fourteen to forty-two days. This variation occurs not only from woman to woman but also from cycle to cycle for the same woman. Moreover, the part of the cycle that varies most is the part that would be most useful to the male – the number of days from the beginning of menstruation to ovulation. Far from being a predictable fourteen days, this phase can vary in length from about four to twenty-eight days in any normal, healthy woman. Neither the man nor the woman can predict the most fertile day of her cycle simply by counting forward from the beginning of her previous period.

Of course, confusing the partner requires a female to do more than vary the day of ovulation and avoid developing a tell-tale crusty bum and red vulval lips. Even without these attributes, a woman would still give the game away if she showed an interest in intercourse only when she was most fertile. She avoids this danger through a sophisticated veneer of subconscious changes in mood and behaviour. First, her body is prepared to allow her partner to inseminate her at any time during her menstrual cycle, both when fertile and when infertile. Secondly, her body shows an erratic succession of genuine, false and take-it-or-leave-it interests in sex throughout her cycle. If she does show a day or two of sexual interest when she is most fertile, it is well hidden among decoy phases of interest interspersed with genuine periods of coolness. Finally, and most sophisticated of all, she confuses her partner well because she also confuses herself. It is no accident that a woman is not naturally conscious of when she is most fertile. The uncoupling of her conscious mind from her body's fertility is as important a part of her body's strategy as all the other elements.

In the face of such a powerful and effective female strategy, the man has no chance of being able to predict the best time to inseminate. As a result, the only subconscious strategy open to him is to try to maintain a continuous sperm presence in his partner. Hence the advantage of routine sex to him as well as to her. If a man manages to routinely inseminate his partner about every two or three days, he should always have fertile sperm inside her – in which case, his chances of fertilising her egg in any given month will be about one in three. One missed insemination, however, could be critical, and in Scene 2 it was. The man failed to fertilise his partner's egg.

When he ejaculated inside her that Saturday, it was a week since he had last done so, and his last sperm would have lost their fertility on Wednesday. His partner ovulated on Thursday night and, although a few sperm had still been alive inside her on Friday, when her egg was still alive, those sperm had been infertile. She did not conceive, and in two weeks' time her next period will begin. Most likely she will begin to bleed on the Saturday of that week. We can predict this with some certainty because, unlike the number of days from the beginning of one period to ovulation, the number of days from ovulation to the beginning of the next period is a fairly predictable fourteen days, varying only from about thirteen to sixteen.

When she does begin to bleed, the couple will almost certainly see their failure to conceive this month as a joint one. However, there is an alternative interpretation – that the woman's body actually engineered the situation to avoid conception this month, at least via her partner.

We saw earlier that women confuse men by seeking or allowing intercourse erratically throughout their cycle. But this is not the whole truth. The incidence even of routine sex does vary a little during the menstrual cycle.

First, both women and men are less disposed to have sex while the woman is menstruating. Some human cultures even have a taboo against menstrual intercourse. We find the same reduction in rate of intercourse during menstruation in all primates that mate throughout the menstrual cycle – even in those, like marmosets, in which no blood flows to the outside. This reduction in sexual interest during menstruation should not be surprising, because menstruation is a time of slightly greater risk of infection to both males and females if they have penetrative sex.

Secondly, and perhaps this *is* surprising, a woman is more likely to have routine sex in the two weeks after she has ovulated, when she cannot conceive, than in the two or so weeks before, when she might conceive. The difference can be detected statistically, but is too slight to be noticeable to either the man or the woman. This subtle change in behaviour is not due to a conscious decision by couples to avoid conception. Women using reliable contraceptives, including the pill, show the same slight changes in behaviour. So, too, do other primates. The subtle change in a female's willingness to have sex with her partner at different stages of her cycle is hormonal, not cerebral.

Engineering when and how often they have sex is only one of the ways in which a woman's body manipulates the chances of her partner fertilising her egg. Another way is to get rid of some or all of his sperm. Most people will never have looked at the damp patch on their sheet after sex and marvelled at women's power and sophistication. My hope, however, is that after reading the next scene, that damp patch will never seem the same again.

SCENE 3

The Wet Sheet

The woman stirred in her post-coital sleep as a familiar sensation began to tickle her buttocks. She opened her eyes and looked at the luminous red numbers on her bedside clock. It was nearly forty-five minutes since her partner had ejaculated inside her. Now she felt the first hint of wetness emerging from her vagina. As she hovered between sleep and consciousness, she tried to decide whether to get up and go to the toilet, to reach for a tissue, or simply to let the familiar liquid ooze out of her vagina, dribble down between her buttocks, and wet the sheet.

During half-consciousness, her mind drifted back seven years to her first semester as a student. At the beginning of her final long schoolgirl summer, she had met a male college student, two years older than her. Within days of meeting they had sex and thereafter did so whenever the opportunity presented itself. At first they had used condoms, but eventually she had agreed to go on the pill. When summer was over and they had each gone to their separate colleges, they had, for a few months, continued their relationship, taking it in turns at weekends to visit each other in their tiny flats. On such weekends, they always spent the Sunday afternoon in bed having sex. Invariably, they would stay in bed until the very last moment. Then there would be a mad scramble to dress and get to the station in time to catch the last train. Whenever she was returning from visiting him she could guarantee that no sooner would she have just settled comfortably in her seat than she would begin to feel the fruits of their intercourse seeping out on to her knickers. She would then spend the rest of the journey with a clammy sensation between her legs.

Now, seven years later, lying in bed, she stirred into wakefulness. With great effort she got out of bed, made her way unsteadily to the toilet, switched on the light, sat down and urinated. As she stood up to flush the cistern she looked down into the bowl. There, in the water, were four white, almost spherical globules. It crossed her sleepy mind as she went back to bed that maybe she hadn't yet conceived because she had a problem retaining sperm. However, no sooner had the thought come than it had gone, and within a minute of lying down she was asleep. Tonight, at least, the sheet had remained dry.

~

There are probably few aspects of sex as misunderstood and maligned as the 'flowback', that collection of material that flows back out of the vagina sometime after intercourse. To most people it is an irritation and to some even a worry, a threat to their fertility.

The flowback is a joint man–woman production. The main part is the seminal fluid introduced by the man, almost all of which is ejected from the vagina. To this the woman adds a quantity of mucus from her cervix. There are also cells from the inside of the vagina, dislodged by the thrusting of the penis. But the most common cells in the flowback are the sperm – and there are usually millions of them. From the human perspective, it is difficult to see the flowback as anything other than a negative, passive event. At first sight it might seem an impossible transformation of image to convert a damp patch on the sheet or a dribble down the leg into a positive, dynamic event. Yet that is just what I want to try to do. I want to argue that the flowback is one of a woman's major weapons in her pursuit of reproductive success.

One of my favourite photographs of recent years is of a family of zebra: a stallion, a mare and a young foal. The stallion has just inseminated the mare and is still standing on his hind legs, his front legs on her back. The foal is looking the other way, seemingly embarrassed to watch, as its mother ejects from her vagina a dramatic gush of flowback. In zebra, within minutes of insemination, the mare promptly ejects a major part of the stallion's ejaculate. Women are not as blatant as female zebra, and it might seem that a dribble down the leg can hardly compare with the zebra's vigorous response. But as part of my research I have had to take a closer interest in flowbacks than perhaps most people, and I can report that women need not, actually, feel at all inferior in this respect.

The woman in Scene 3 noticed white globules in the toilet after she had urinated. If you are female, try using a mirror to watch the flowback emerge when you urinate about thirty to forty-five minutes after intercourse. You can't do this on a toilet, so instead try urinating into an empty bath. Crouch down. Separate your pubic hairs and vulval lips so that the urine will squirt forward. Choose your moment. Wait until you can feel that the flowback is gathering, then urinate. Viewing it from the side you will see that the urine stream shoots forward out of the urethra while, a centimetre or so lower down, when the correct muscles contract, the flowback is squirted out of the vagina with impressive force. (If you are male, see if you can persuade your partner to let you watch her eject the flowback.) Whether you are male or female, you will be left in very little doubt that the flowback is the female's *ejection* of part of the inseminate that she has just received from the male.

Women and zebra are not the only two female animals able to do this. Monkeys, rabbits, mice, sparrows – and probably all other female mammals and birds – also eject a flowback.

How do females do it? Before I can explain what happens in humans, I need to do two things. First, I need to describe in some detail the architecture of a woman's reproductive tract. Secondly, I need to describe what happens to the ejaculate in that critical first half-hour or so after insemination. These descriptions will take some time.

Imagine that you are a doctor and that you are just about to give the woman on the bed in front of you an internal examination. She is lying on her back. First, you part any pubic hair that is in the way and separate the main lips so that you can see the entrance to her vagina. The chamber just inside is the vestibule. If your eyesight is good and you separate the vulval lips well enough you will see, opening into the top of the vestibule, the urethra through which she urinates.

Next, you slip two fingers between the vulval lips into her vagina and gently push them as far as they will go. First, note that the vagina is in contact with your fingers all the way round. This is because, when there is nothing inside it, the vagina is not a tunnel but a slit, with the two walls pressed together. And not only is the vagina not a tunnel, it is not even a throughway. The popular image of it as a tube leading straight into the womb through the cervix is quite wrong. It is also quite wrong to imagine that dead-eye dicks can actually shoot their ejaculate through the cervix straight into the womb. Both of these images are false because the vagina is, in fact, a dead-end. Of course, there *is* an exit into the womb, but it is not straight ahead; to find that exit requires virtually a right-angled turn.

Without withdrawing your fingers, turn your hand so that its back is on the bed and your palm is facing upwards. The womb, which is pear-shaped, is balanced on top of the far end of the vagina, probably just beyond your fingertips. The narrow end of the pear is the cervix and it is the cervix that penetrates the roof of the vagina, projecting through by a couple of centimetres. If your fingers are long enough – many aren't – their tips can feel the cervix sticking through the vaginal roof. The cervix has a narrow channel running though it, and it is this channel which connects the vagina to the inside of the womb and through which the sperm must pass on their way in. It is also through this channel that, in a phenomenal feat of engineering and elasticity, a baby must pass on its way out. For the moment, though, let's concentrate on the narrowness of the channel and on sperm going in.

The channel through the cervix is not empty. It is filled with mucus and, if you leave your fingers inside this woman long enough, some cervical mucus will flow out on to them. This is the woman's main contribution to the flowback and it has a starring role in this book. To understand much of human sexuality, we need to appreciate the beauty of a woman's mucus and the amazing things she does with it. She has complex requirements of her cervical mucus. On the one hand, it is her last defence against the bacteria and other disease organisms which are forever trying to invade her cervix and womb. On the other hand, she needs it to allow passage to sperm on their way in and to her menstrual flow on its way out. In other words, she needs it to function as a two-way filter.

Most people think of mucus rather contemptuously as a messy, amorphous substance, probably because their main contact is with the mucus that comes out of their noses. Cervical mucus may look and feel like nose mucus, but it is in fact very different. It is wonderful stuff with an immaculate structure and is absolutely vital to a woman's health, safety – and sexual power. It contains fibres and is permeated by channels. Most of these channels are very narrow, some only the width of two sperm heads side by side, but they are none the less the highways through which sperm swim as they migrate from the vagina to the inner regions of the cervix and beyond.

Cervical mucus is secreted continuously, primarily by glands in the top half of the cervix, furthest from the vagina. After being secreted, it slowly flows in glacier-like fashion down through the cervix, eventually dripping into the vagina. The rate of flow of this cervical glacier is slow compared with the speed of a swimming sperm but fast compared with the speed of invading disease organisms. Bacterial and other invaders are carried out of the cervix back into the vagina before they can take hold. In the vagina, they are killed by the acidity of the vaginal juices. During menstruation, the menstrual flow simply adds to the mucus flow. The double flow makes it even more difficult for disease organisms to invade – especially important because during menstruation the raw lining to the womb makes it particularly vulnerable.

The demands that women make on their cervical mucus are particularly great because of their strategy of copulating when they would seem to have little use for sperm. We have already discussed why women have sex at infertile stages of their menstrual cycle (Scene 2) – to confuse men. Post-menopausal women also copulate to confuse men, often continuing to be sexually active for many years after their last period. By preventing a partner from identifying the end of her reproductive life with certainty, a post-menopausal woman is able to reduce the chances of being deserted for a younger, more fertile woman. In fact, women who are apparently post-menopausal can occasionally conceive – at least up to the age of fifty-seven and reportedly up to the age of seventy (Scene 34). Even pregnant women continue to copulate – again to confuse men, but for particular reasons that we shall discuss in connection with Scene 17.

At all times, a woman has to balance the advantages of letting sperm through and keeping disease organisms out. Clearly, making life easier for sperm makes it easier for disease organisms too. During pregnancy, a woman has no use for the sperm she collects during copulation, and she ejects the whole inseminate in her flowback; her cervical filter makes life impossible for sperm in order to maximise her defence against disease. At all times other than during pregnancy, however, a sexually active woman may have some use for sperm, and she then needs to sacrifice some of her defence against disease in order to allow sperm through. And just as the advantage of allowing sperm passage varies over the span of her life and during her menstrual cycle, so too does the strength of her cervical filter.

A non-pregnant woman has least use for sperm during the sub-fertile phases of her life (such as during most of each menstrual cycle and after the menopause). Even during these phases, though, she gains some benefit from allowing sperm passage, because sperm allowed through during a sub-fertile phase can influence the sperm present at the beginning of any next fertile phase (as we shall first see in Scene 7). But since the advantage of retaining sperm during sub-fertile phases is relatively small, a woman can afford to make her cervical filter more hostile to sperm in order to increase her defence against infection. At the approach of ovulation, the advantage of allowing more sperm through her cervix naturally increases, so she makes things as easy as possible for sperm at this time. The way she facilitates or impedes sperm passage during fertile and sub-fertile phases is by altering the nature of the cervical mucus.

Throughout a woman's many and long infertile phases, cervical mucus is made difficult to penetrate. The narrow mucus channels are small in number, and although sperm can enter the mucus, few can swim through it. Even those that *can* penetrate do so more slowly. During this phase the flow of mucus is slow, but fast enough to do its job of combating disease. In contrast, during her short fertile phases, the mucus changes: it becomes much more liquid and stretchy and the channels become bigger. It is more easily invaded by both sperm and bacteria.

The only major problem faced by invading sperm during a woman's fertile phases is that not all channels are clear of the blockages referred to earlier. To remove these, and to combat the increased risk of infection, the woman increases the flow rate of her mucus. This way she flushes out cells, bacteria and other debris. She is aware of being wet more of the time and a clear, sweet-smelling secretion appears on her underwear.

Although the benefit of these changes in the cervical mucus is clear, they could create a problem. They *could* threaten a woman's attempts to hide her fertile phase from both her partner and herself (Scene 2). Her body overcomes this threat by making the increase in mucus secretion more erratic and more spread out than would be necessary just to aid the passage of sperm through her cervix. The mucus symptoms can occur more than a week before ovulation and can continue for two or more days afterwards. Consequently, although cervical mucus gives some clue as to the timing of a woman's fertile phase, it is too unpredictable to ruin her overall strategy.

So, cervical mucus is an adequate sperm filter in its own right. And no matter what the phase of her menstrual cycle, a woman can enhance the mucus's filtering efficiency by blocking the channels. The more channels she blocks, the stronger the filter. So what does she use to block her mucus channels? There are three things. One is the blood, tissue and other debris from menstruation. Another is white blood cells (Scene 4). And the third is sperm (Scene 7). These blocks may last for several days but are eventually lost when they are carried inexorably by the cervical glacier of mucus into the vagina. Later, we shall see that this ability of a woman to enhance, or not to enhance, her cervical filter is a most powerful weapon in her attempts to outsmart men (Scenes 22 to 26).

Even once it is in the vagina, the mucus's job is not finished. It flows down the walls of the vagina, coating them with a thin film. Some exits, contributing to the 'wetness' a woman feels on her vaginal lips. Much of the mucus film, however, remains on her vaginal walls – representing advance preparation for her next intercourse, even if it does not happen for days. When she eventually becomes aroused during foreplay, her vaginal walls begin to 'sweat'. The sweat itself is not slippery. But when it mixes with the film of old cervical mucus, the result is a very effective lubricant. The vagina is now ready for penetration and intercourse.

We now have all the information necessary to follow the events that take place from first penetration of the vagina by a penis to the production of the flowback. But to help us along, we require a change of image from that of the internal medical examination that we have used so far. What I am about to describe was first filmed by strapping a fibre-optic endoscope to the underside of a man's penis just before he and his partner had sex. This gave a penis's-eye view of what happened; so, to help me in my description, suppose that you have volunteered to take part in such an experiment. You are having intercourse in the missionary position and your erect penis (if you are male) or your partner's (if you are female) has a camera on its tip. You can see what is being filmed on a big TV screen on the wall in front of you.

As the penis pushes forward into the vagina for the first time, the vagina walls part and, when the penis is fully in, you can see the blind end of the vagina some distance ahead. Still slightly ahead, sticking through the vagina's roof, is the cervix. At the moment, with its central, dimple-like opening, it looks like a pink sea anemone shorn of its tentacles. But it will change as intercourse proceeds.

If you watch the screen when thrusting begins, you will see that each time the penis pulls back the vagina walls close behind it. Each time the penis pushes forward, the walls part. Whenever the penis is fully inserted, you can see the end wall of the vagina and the protruding cervix. As thrusting continues, the picture at full insertion changes. The far end of the vagina becomes more like a chamber, slowly filling with air and becoming slippery with mucus. Even more dramatically, the cervix begins to stretch and hang down more and more. Gradually it looks less and less like a sea anemone and more and more like a pink, rather broad, elephant's trunk. Eventually, all you see in front of the fully inserted penis is the front wall of the cervical trunk. Its opening points down to the vaginal floor and cannot really be seen. Towards the climax of intercourse, the cervical opening may even rest on the vaginal floor. When the penis ejaculates, the spurts of semen hit the front wall of the cervix and run down on to the floor of the vagina, forming a pool at the bottom of the chamber. Hanging down, dipped into this pool of semen, for all the world like an elephant's trunk at a watering hole, is the cervix.

After a minute or so, with ejaculation complete, the penis begins to shrink. As it does so, the vaginal walls close behind it, helping it out but keeping the pool of semen in. With the shrinking of the penis, we lose our camera-bearer and our TV screen goes dark. However, by now it doesn't matter much – although critical events are taking place, these are chemical and microscopic rather than obvious.

The first thing that happens – and we might almost have seen this on our TV screen just before the penis shrank too far down the vagina – is that the seminal pool coagulates, becoming slightly less watery and slightly more jelly-like. The second thing that happens is that sperm start to migrate out of the seminal pool. Their destination is the cervical channel, which they can enter only by passing through the interface which forms between the cervical mucus and the semen. Imagine that the cervix really is an elephant's trunk, dipping into a large pool of semen. The trunk is full of mucus. This mucus, however, does not dissolve in or even mix with the semen when the two come into contact. Instead, something much more structured takes place.

The interface between mucus and semen at the mouth of the cervical trunk is not flat. 'Fingers' of semen enter some of the larger channels in the cervical mucus, and grow. They penetrate a short distance into the cervical trunk, stretching upwards into the mucus like the fingers of a rubber glove. Sperm frantically swim into these fingers, and from there stream into the narrower channels of the mucus, leaving the seminal fluid behind. Later we shall follow these sperm further, but for the moment our interest is in the flowback.

After a few minutes dipped into the seminal pool, the cervical trunk begins to shrink back to the roof of the vagina, metamorphosing once more from elephant's trunk to sea anemone. It loses contact with the seminal pool and hence cuts off the upwards escape route for the sperm. Once the cervix has withdrawn, those sperm still in the pool are condemned to ejection and an early death. About fifteen minutes after ejaculation, the pool begins to decoagulate and becomes more watery again. Soon, imperceptible and unconscious muscular ripples begin to massage the mixture of old semen, mucus, sperm and other cells down the vagina. Eventually, the mixture collects in the vestibule. On average, this happens about half an hour after ejaculation, but can occur as soon as ten minutes or as long as two hours. Before this, a woman can stand up, walk around, even urinate, and she will not eject the flowback. Once the flowback has collected in the vestibule, however, any one of these activities, or even a cough or sneeze, will rid her of the unwanted material. Even if she stays asleep, the flowback becomes so liquid after about two hours that it will eventually begin to seep out anyway, producing the wet sheet.

On average, the flowback contains about half of the sperm that have been introduced – sometimes more, sometimes less. How many depends in part on the severity of the woman's cervical filter. Quite often (about one in ten occasions) the filter is so severe that she ejects almost all of the inseminate; more rarely, her filter is so weak that she keeps nearly all. Most important of all, the proportion of sperm that she keeps is not due to simple chance. To a large extent, it is under the control of her body – and not just her cervical filter. Each time she has sex, a woman's body decides how many sperm to keep and how many to eject. How and why, we shall see later. It won't be long before this female ability becomes very important in the lives of our couple. But not just yet.

SCENE 4

Topping Up

Over the next two weeks the couple became quite active, sexually. The coolness of the woman's fertile week evaporated. Both partners went through a phase of anticipating and enjoying their sexual activity more than they had for about a year. After making up on the Saturday night, they had sex twice on the Sunday, once in the morning when they first woke and again in the afternoon at about three o'clock. Half an hour later, they even tried again. He had an impressive erection but, despite ten minutes of intermittent thrusting and encouragement, he eventually had to accept that he wasn't going to ejaculate. Then they missed a few days. Wednesday night was the woman's weekly night out with girlfriends; Thursday night the man's weekly night out 'with the boys'. On both nights, when the reveller eventually crawled into bed, the partner was asleep, or at least pretending to be. On Friday night, however, they had sex and did so again on both Saturday and Sunday. The next week followed a similar pattern, until the woman's period began on the Saturday morning. Then they abstained until the following Saturday, by which time her menstrual bleeding had finished.

~

Few couples have their routine sex at absolutely fixed intervals. In the four weeks we have been following this couple, they have had penetrative sex ten times and the woman has been inseminated nine times. But the time interval from one intercourse to the next has varied from as short as thirty minutes (albeit without ejaculation) or seven hours (with ejaculation) to as long as seven days.

Men get a fairly rough deal in this book. Our story will be one of men's bodies forever trying to make the best of a bad job, while the woman's body outsmarts and outmanoeuvres them at almost every turn. But this at first sight uninspiring scene does give us an opportunity to watch men doing something fairly impressive. A man may not *look* particularly sophisticated at the moment of ejaculation, but something remarkable is, in fact, taking place. Each time he has intercourse during routine sex, he introduces no more sperm than are needed to 'top up' his partner. How does such restraint help him in his pursuit of reproductive success? To understand what the man is trying to do, we have to follow further those sperm we last saw swimming through the channels in the woman's cervical mucus.

A small proportion of these sperm, the vanguard, swim straight through her cervix into the womb. Except when she is pregnant, her womb is roughly pear-sized as well as pear-shaped. As with the vagina, the walls press closely together, so there is little space inside. Once in the womb, sperm swim close to the walls and are helped by the womb to reach its top, the widest part of the pear: in effect, they surf-board, carried on the crests of muscular ripples passing along the womb's walls. At the top of the womb on each side (where horns would be if the pear shape of the womb were a bull's face) is the opening to a narrow tube, the oviduct. Although there are two oviducts, only one will contain an egg during any given menstrual cycle. Once out of the womb, the sperm swim a short distance along an oviduct until they reach a rest area. Here they cease swimming, settle down, and await developments.

Back in the cervical mucus, another set of sperm swim along more diagonal channels and stream into tiny crypts in the wall of the cervix. These sperm also, once in the crypts, cease swimming, settle down, and conserve their energy. Over the next four to five days they will gradually wake up and re-enter the cervical channel. Then they, too, will complete their journey through the mucus, surf-board through the womb, and head for the rest area in the oviducts.

The final set of sperm simply stay in the cervical mucus. They just sit there, cluttering up the mucus channels. Eventually they die – or are killed. Their lethal assailants are the marauding hordes of white blood cells which are unleashed by the female from the walls of her womb within minutes of insemination. As they advance through the cervical mucus, these killer cells engulf and digest live and dead sperm alike. At their peak the white blood cells can match the numbers of sperm, but within twenty-four hours of insemination these hordes have gone, leaving behind much smaller numbers to complete the mopping-up operation. Multitudinous though the white cells may be, they do not pursue sperm into the cervical crypts.

An average inseminate contains about three hundred million sperm. Of these, the woman will eject about 150 million in the flowback. A few hundred sperm may go straight to the oviducts and about a million may first go into the cervical crypts to form reservoirs, leaving in order to complete their journey to the oviducts over the next five days. In all, about twenty thousand sperm from each inseminate eventually pass through the oviducts. The remainder, those not ejected in the flowback, clutter up the cervical mucus, eventually to be mopped up by white cells or carried by the slow, glacier-like flow of the cervical mucus (Scene 3) back into the vagina.

It might seem exceedingly wasteful to introduce three hundred million sperm when only about a million enter reservoirs. But all is not what it seems. As far as topping up the woman is concerned, the important point is that the size of the reservoirs depends on how many sperm the man introduces. If he puts in only two hundred million, then the reservoirs will be only half as full as if he put in four hundred million.

From every insemination, a man and woman contrive between them to produce a steady passage of fresh sperm through each oviduct over a period of about five days. This traffic probably peaks about one to two days after insemination, then gradually declines as the reservoirs of sperm in the cervical crypts slowly shrink. This is where topping up comes in. If the male can keep the reservoirs of sperm in the cervical crypts topped up, he can ensure a continuous passage of fresh sperm to the rest areas in the oviducts. If he injects more sperm than are needed, they will be wasted; the reservoirs will overflow and even more sperm will simply hang around in the cervical mucus and fall prey to the female's white cells. There is also a danger that too many sperm will arrive in the oviduct and, with a surfeit of the chemicals that they carry on their heads (Scene 7), actually kill any egg that may be present. On the other hand, if the man doesn't fill the reservoirs enough, too few sperm will arrive in the oviduct or the reservoirs will dry up prematurely. The man's challenge is to adjust the number of sperm he introduces according to the number needed to top up his partner's reservoirs. This he seems to do with remarkable precision.

The adjustments we are talking about are roughly as follows. If more than a week has passed since the man last inseminated his partner, her reservoirs will be empty and he will introduce a full load of sperm, say four hundred million. Of these, perhaps just over a million will fill the reservoirs. If the gap is only three days, he will introduce about two hundred million, and half a million will top up the half-empty reservoirs. If the gap is only three hours, he will introduce about thirty million and if only a few minutes, he won't introduce any. Even after half an hour, he may find it difficult to ejaculate, like the man in our scene. His body is effectively saying that there is no point. His partner's reservoirs are full and any sperm ejaculated will simply be wasted.

Our couple have had penetrative sex ten times in the four weeks we have watched them. In that time, about three thousand million sperm will have passed from one to the other. So precise is the man's ability to top up his partner that if they had doubled their rate of intercourse, or if they had halved it, it would have made scarcely any difference to the total number of sperm she would have received.

So men's bodies inject only the number of sperm needed to top up their partner. How do they do it? To answer this, we need to understand more about both the architecture of the male reproductive tract and the mechanics of ejaculation.

Imagine that you are a doctor, sitting on a chair, and a naked man is standing in front of you waiting to be examined. His genitals are at your eye level. Note his navel, his pubic hair, and the penis hanging down, slightly askew, in front of the scrotal sac containing his pair of testes. Hold his flaccid penis in the palm of your right hand and, if he has one, make sure the foreskin is pushed back. The rather swollen knob at the end of the penis is called the glans and, right in front of you, is the vertical, slit-like opening of the urethra. Through this, the man both urinates and ejaculates. Fix the line of his urethra in your mind. It runs in a straight line from the opening slit, back through the penis shaft, into his body and up to join his bladder. Fix your eyes at the top of his pubic hair and imagine you can see inside to where the urethra and the bladder join. Just below this point, the urethra is joined from left and right by two tubes. These tubes run all the way down to the testes and each one contains, in effect, a column of sperm. Where these tubes join the urethra, they are surrounded by a walnut-sized mass of tissue. This is the prostate gland, which produces the bulk of the seminal fluid.

So where do these two columns of sperm come from? Even as the man stands in front of you, his testes are a hive of activity. Inside, cells are multiplying, growing and finally maturing into sperm. By the time the sperm are mature and fit to be ejaculated, they have already been herded into the single column of sperm from their testis. They are in the sperm tube, but they are still in, or rather on the surface of, the testis. The sperm tube changes in character along its length: in the testis the tube is called the epididymis, and from the testis up to the urethra it is called the vas deferens. Whereas the vas deferens is more or less straight, the epididymis is incredibly zigzag and convoluted.

Once the sperm are in the epididymis they are, in effect, simply queuing up to be ejaculated. Each time the man ejaculates some of his sperm, the rest shunt forward. As part of the front of each queue is lost through ejaculation, newly mature sperm join the back of the queue in the testis. Very approximately, it takes two months for a sperm to develop and travel from deep inside the testis and join this queue. Each sperm will then spend a further two weeks queuing in the epididymis and up to a further five days or so in the vas deferens. There is a little queue-jumping as some young sperm from the back of the queue are shunted ahead of older sperm at the front, but this need not concern us here.

Now, let's follow what happens to our medical model as he goes off with his two columns of sperm and has intercourse. While he was standing in front of you he had no sperm in his urethra, but two sperm tubes containing in total up to about a thousand million sperm. Nothing will change when his penis becomes erect, not even during the early stages of penetration and thrusting. Eventually, though, sperm will be shunted out of each sperm tube and into the urethra. A round sphincter that normally prevents urine leaking out of the bladder also prevents sperm from entering the bladder. The man's urethra is now loaded, ready to fire.

While loading, the man will feel a pleasurable urgency at the base of his penis. He will also know that ejaculation is imminent. Just how imminent is, to a limited extent, under his conscious control. When he finally ejaculates, seminal fluid pours from his prostate into his urethra. Then muscles contract and the mixture of fluid and sperm is projected in a series of spurts along the urethra and out into the woman.

Now, it is not difficult to understand how the man's body controls the number of sperm he ejaculates. Depending on how many of his loading muscles work, and how strongly, he can shunt any length of each of his sperm columns out of their tubes and into his urethra. Even after loading, his body can change its mind. By varying the number of spurts, usually from between three and eight, he can ejaculate a different proportion of the loaded sperm. Any sperm and seminal fluid left in his urethra after ejaculation can be flushed out at the next urination.

Of course, somewhere there needs to be a link between the man's brain, which keeps track of when he last inseminated his partner, and his genital musculature. Given such a link, it is not difficult to see how his body can top up his partner so accurately. Nobody suggests, of course, that a man has *conscious* control over sperm number. In the midst of thrusting and at the point of loading his urethra, he does not ask himself consciously, 'Is this a hundred-million occasion or a four-hundred-million?' His subconscious mind and body have done this for him. When the moments to load and ejaculate arrive, his various body parts respond accordingly. This leaves his conscious mind free to concentrate on his thrusting – and the woman.

SCENE 5

Conception

It is Friday night and twenty-one days have passed since the beginning of her last period – bleeding that signalled a second month without conception. Briefly, the couple had feared for their fertility.

They were calmed, however, both by the experiences of friends and by the discovery that a couple must have unprotected sex for a whole year before they qualify for medical investigation. With these reassurances, they had put the month behind them and started afresh. Now, they have just had sex and are drifting off into post-coital sleep. Tonight, she won't wake to go to the toilet and her flowback will wet the sheet.

In the past two weeks, their sex life had more or less followed the routine they had established since giving up contraception. Sex on Saturday, Sunday and sometimes, as this week, also on Friday had become the norm. This week, though, had been slightly different. On Wednesday night, as soon as she had got into bed after returning from her weekly night out, she had stroked his body until he awoke and had played with his penis until he was erect. Then, she had sat astride him and manhandled his penis into her vagina. He had slipped in easily, because she was very wet, and from that moment she had done all the work. Slowly he had begun to enjoy himself. They weren't very practised at having sex with her on top, and a couple of times he had slipped out. She had had to work quite hard but eventually he had ejaculated. Tonight, they had tried the same position but this time, for some reason, it just hadn't worked. Eventually, they had reverted to their usual missionary routine.

As she drifted into sleep, things were happening inside her body that would change her life for ever. She had ovulated earlier that evening and her egg was just reaching the place in her left oviduct where fertilisation would occur. As it arrived in the fertilisation zone, three fertile sperm arrived simultaneously. They began to burrow through the outer layers of the egg. Two of them were delayed for a few seconds as they bumped into each other, attempting to penetrate the same point in the egg's defences. The prize of fertilisation went to the third sperm, which had a clear run, unhindered by any others. By the time another sperm arrived only a few seconds later, the egg had put up its barriers and there was no way in. The egg had been fertilised by the first sperm to get through. Three months after giving up contraception, the woman had conceived.

In twenty days' time, an overdue period would prompt her to carry out a pregnancy test. Two hundred and fifty days after that she would give birth. The identity of the father, however, would never be known. For the sperm waiting in her oviduct that Friday night had in fact been from two different men.

~

There is one final phase to the odyssey of fertilisation – a phase that can be critical in a man's and a woman's pursuit of reproductive success.

In previous scenes, we have followed sperm from their earliest life in the testes to the moment they are shot into the woman's vagina. We have watched them attempt to escape from the seminal pool, migrate through the narrow channels of cervical mucus and eventually make their way to a rest area in an oviduct. There we left them, waiting for the final stage of their journey.

Sperm may rest in an oviduct for as long as a day and, at any one time, up to a few thousand may be resting and waiting. One by one, they wake up and swim further along the oviduct. Their initial destination is a region in which, if an egg is present, fertilisation can occur. But usually no egg is present, and the sperm simply pass through and eventually die.

As a sperm approaches the fertilisation zone, it changes its behaviour. Its tail starts to beat more vigorously and, once it arrives, it often swims frantically in circles or figures of eight. At any one time, the fertilisation zone may contain anything from one or two to up to a thousand sperm, all looking for the egg that usually never comes. After a while, they leave the zone, one by one. As each leaves, its place is taken by a new, fresher arrival from the rest area. Having left the fertilisation zone, each sperm then travels the remaining distance up the oviduct until eventually it swims out into the woman's body cavity – which it can easily manage because the end of her oviduct gapes open, its opening surrounded by finger-like projections.

Now let's consider the egg.

A short distance away from the end of each oviduct, suspended like a relatively huge planet next to a black hole, is an ovary. Tiny hairs on the inside of the oviduct create a current in the body fluid so that when an egg is released by an ovary, it is slowly wafted towards the black hole of the oviduct. Like a waiting hand, the finger-like projections funnel the egg into the tube. From here, the egg begins its five-day journey down towards the womb.

To achieve fertilisation, a sperm has to do more than simply encounter an egg in the fertilisation zone, because the egg arrives in that zone surrounded by three lines of defence – a fortress that has to be breached by the sperm before the egg will yield. The outer line of defence, the cumulus, is a thick layer of shapeless cells which the egg has brought with it from the ovary. Beneath the cumulus is another relatively thick, smooth layer, the zona, which is the outer membrane of the egg itself. Under the zona is a narrow space which surrounds the final, most vulnerable barrier, the vitelline membrane.

Using its head, the sperm hacks its way through the cumulus cells. If successful, and it reaches the zona beneath, it sticks the side of its head on to the membrane with chemicals. With this chemical attachment serving as an initial purchase, the sperm again uses its head to cut a way through, this time employing a pointed spike that has been exposed at the tip of its head. The lashing tail provides the force to push the sperm forwards. Finally, if this sperm is the first to get through the zona, cross the underlying space and touch the vitelline membrane, it is engulfed by the egg in a welcoming embrace. Having embraced one sperm, the egg passes a chemical message across its surface and, within seconds, becomes impenetrable. If you are a human sperm, there is no prize for coming second.

The successful sperm sheds its membranes within the egg, releasing its genetic heart of DNA. This then travels to fuse with the similar heart of the egg. Fusion of the DNA from sperm and egg mixes together the genes from father and mother in equal proportions. A new person has been conceived, with characteristics that are a subtle mixture of his or her two parents.

We know with certainty the mother of the child in our scene. After all, the child will spend nine months developing inside her. But who is the father? As we saw, in those few critical days before she ovulated the woman had collected sperm from two men – her partner and her lover. To see what happened, and to work out who is the father, we need first to go back ten days. We are about to witness a sperm war.

3

Sperm Wars

SCENE 6

A Chance Affair

The woman was out on her Wednesday night ritual with eight of her girlfriends. She had been going out like this for over a year. About a dozen of them were involved, but not everybody went out every week. Typically, their night centred on lots of drinking and talking, a meal and sometimes a club. Occasionally, a man or two would move in and try to prise somebody away from the group. They expected each other from time to time to talk to, and perhaps even leave with, a man. Even though most of them had a male partner at home, the group enjoyed an unspoken sisterly complicity.

Tonight, it was her turn. Quite by chance, who should walk into the bar but the man she had met during her last schoolgirl summer – the one who, for a few months, had been her weekend sexual partner during her first term as a student. They recognised each other immediately and spent virtually the whole evening talking, catching up on events since their last, rather acrimonious, meeting. She learned that he now worked at the other end of the country but was in town for a week on business, staying in a nearby hotel. Now nearly thirty, he was still not living with anybody, but had a girlfriend – or two.

He still oozed masculinity, promiscuity – and unreliability. Kudos though there had been in being his girlfriend, she had eventually finished with him on discovering his many and varied infidelities. At that vulnerable stage in her life she had needed someone she could rely on. Now, however, meeting him out of the blue, many of her old feelings returned. All the same, at the end of the evening she rejoined her group.

At lunchtime the next day, he turned up at her work and took her out for a snatched bar lunch. Over lunch, they arranged to meet for a meal that evening. As it was Thursday and her partner would be out with his friends, she reasoned there was no need or point in telling him. As far as she was concerned, the evening would be totally innocent and therefore not worth mentioning anyway. Even so, she took her ex-boyfriend to a restaurant some way out of town where they were unlikely to be seen.

All through the evening, it was obvious he was expecting them to end up in bed back at his hotel. He was very attentive and flirtatious and from time to time would find an excuse to touch her. But the thought of infidelity never seriously crossed her mind. She still found him attractive and his touches arousing, but she also found his expectation of sex irritating, almost offensive. As a result, she had long cool phases towards him throughout the evening. Eventually, he got the message and backed off. On the return journey to her home they simply exchanged pleasantries.

In the few moments before she got out of his car, they spoke as if they were never going to meet again. She surprised herself when a sudden feeling of warmth, nostalgia and maybe guilt prompted her to kiss him briefly on the cheek. She surprised herself even more when she kissed him again, this time on the lips. Somewhat flustered by this momentary surge of passion, she got quickly out of the car, wished him a good life, and went inside.

When her partner returned home an hour later, she was in bed pretending to be asleep. As he fell into a drunken sleep and began a night of snoring, her thoughts and dreams raced around with the excitement of the evening. At some unknown hour of the night she surfaced just enough to realise she had climaxed in her sleep.

All the next day, at work, she couldn't believe how easy it had all been. She had spent the evening, albeit quite innocently, with another man, and absolutely no one had known about it. The man, their evening, their conversation and their kiss, as well as memories of their earlier sex life when she was still a teenager, were at the back of her mind continually as she worked. Her thoughts kept her in an almost permanent state of subliminal excitement. Her knickers were damp virtually all day and on one visit to the toilet she masturbated.

She didn't have sex with her partner that Friday night but she did both on Saturday and on Sunday. On Saturday night she more or less insisted that he gave her an orgasm before entering her. It was rare for her to climax during intercourse itself. She never expected to and on the occasions when she really wanted an orgasm she made sure it was during foreplay. On Sunday morning she had masturbated while in the bath, then appeared naked in the lounge and seduced her partner into sex on the floor. During each orgasm over the weekend, even during foreplay, the fantasy in her mind was not of her partner (it never was these days!) but of a real or imaginary scene from the past with her ex-boyfriend.

All weekend, she secretly relished her burst of sexual activity and excitement. At no point, however, did she ever contemplate doing anything more dangerous than fantasise about illicit sex. But on Monday, back at work, her mood began to change. Her ex-boyfriend was leaving on Thursday and she might never see him again. An idea began to form in her mind and her excitement slowly turned to nervousness. Maybe she should see him just one more time. It would be easy to arrange. Instead of going out with her friends on Wednesday night, she could spend the evening with him. All she had to do was pick up the phone, call his mobile, and arrange it. Easy.

The thought both excited and frightened her, so much so that all day Monday she did nothing but relish it. On Tuesday she found the courage to phone just once, but got no answer. After that, her nerve failed and she didn't try again that day. On Wednesday morning, her mood changed from fear and guilt to one of calm confidence. Why shouldn't she see him again? He was an old friend and this could be her last opportunity. After all, their previous evening together had been innocent enough – there was no need to feel guilty or nervous. But maybe there was no need to tell anyone, either.

At her third attempt, he answered his phone. He was pleased and surprised to hear from her but was in a hurry and, without time to discuss arrangements, suggested that she call for him at his hotel that evening. She agreed, and spent the rest of the day in a state of high excitement. She told her friends at work that she had to visit her sister, and so would not be with them at their weekly get-together. She told her partner, as she left him at seven o'clock, that she felt like going on to a club and so might be late. He complained, but not much.

She was nervous as she arrived at the hotel and the first few minutes of conversation were very awkward. However, even before they had finished their first drink at the hotel bar, it was as if they were students again. The last six years apart had never happened. Tonight, her mood and behaviour were quite different from the previous Thursday. After another drink she sat with her knees touching his, and as the conversation flowed touched him frequently on his leg or arm with her open hand. When he suggested they eat in the hotel restaurant 'to avoid going out in the cold', she readily agreed. After eating, he 'just had to go up to his room' to get a photograph to show her. She went with him because she had 'always wanted to see what the rooms were like in this hotel'.

He never did show her the photograph. Within moments of closing the door they were kissing and removing each other's clothes. Almost before she could take a breath they were naked on the floor and he was inside her, ejaculating. She had been taken aback by his urgency but made no attempt to slow him down. He did not offer to use a condom or to withdraw, and it never crossed her mind to ask. Her vagina had been wet with anticipation all day and as she walked into his room it had been in full flow. Penetration had been quick and easy and ejaculation had been swift.

When it was over he apologised, saying that his urgency was because he had never stopped loving her and had been desperate for her. He promised that if they got off the floor and into bed he would make it up to her. And he did. For half an hour he caressed and played with her body with an understanding of her woman-hood that her partner had never shown. After she climaxed, they dozed for a while in each other's arms. Then they began all over again. He still penetrated her early, but this time his urgency had gone. Thrusting was slow and long. Unusually for her, she climaxed during thrusting, just a few seconds before he ejaculated.

They settled into another post-coital embrace, but their tranquil-lity did not last for long. For the first time that night, guilt and panic began to grow in her mind. It was getting late. Quickly, her fear became all-consuming. She had to get home. He urged her to stay the night – to phone her partner and give some excuse for not going home. But she would not hear of it. She just had to get home. Eventually, she escaped from his bed on the pretext of going to the toilet. Then she refused to return and began to get dressed. Conversation became tense and forced. She even became irritable with him and their parting was an awkward affair. As she taxied home, her flowback seeped on to her knickers in a replay of past times. But she did not notice. Her mind was too full of what she should do when she got home.

Quietly, so as not to wake her partner, she undressed, washed herself thoroughly, then got into bed. Then she set about waking and arousing him. When he was erect, even before he was fully awake, she sat astride him, inserted his penis, and after a while managed to get him to ejaculate inside her. He dimly registered that she was very wet but thought no more of it, and concentrated instead on enjoying an intercourse without effort.

The next day, her ex-boyfriend left for home. They were never to see each other again. The day after that she ovulated and conceived. Over the next three weeks, she and her partner had sex nearly every other day. By the time she discovered her pregnancy, her night of infidelity had become a distant memory. The guilt and fear had all but gone and she could almost believe it had never happened. She gradually convinced herself that the baby was in fact her partner's. After all, she'd had sex with him about sixteen times that month compared with that single night with her ex-boyfriend.

Nine months later she was to have a daughter. Two years after that, she and her partner had a son, followed three years later by another daughter. As the elder daughter grew she came to look more and more like her mother. She also became noticeably more attractive, more dynamic and more popular than the two younger children. However, the differences between them seemed no greater than between many brothers and sisters.

In the years to come, the woman's partner never once suspected that at the conception of her first child her body had contained sperm from another man, as well as from him. Even she, of course, never knew for certain what happened inside her body over those few critical days. Neither of them was ever to know that the tiny sperm that entered her egg to produce her first daughter was in fact not her partner's but her ex-boyfriend's.

～

The two scenes (6 and 7) in this short chapter explore first the promotion and then the process of sperm warfare. In fact, only one of these scenes involves people – the one we have just witnessed. Scene 7 is not really a scene at all. As a description of sperm wars in action, it is the only occasion in this book in which it was more appropriate to combine scene with interpretation.

We have just witnessed a textbook scene of infidelity. The subtleties of behaviour all play their part in influencing the paternity of the woman's first daughter. We should note in passing the unreliable, promiscuous masculinity of the woman's ex-boyfriend and that his child showed certain more positive characteristics than the children the woman had with her partner. We should also note the number and timing of her orgasms, both male-induced and via self-stimulation. We will discuss later the significance of these elements. Here, we concentrate on the infidelity itself and on the factors which led the ex-boyfriend's sperm, rather than the partner's, to achieve fertilisation. Clearly, the outcome had a major impact on the reproductive success of all three characters in the scene.

We have already mentioned that a woman is slightly more likely to have routine sex with her partner during the infertile post-ovulatory phase of her cycle (Scene 2). The same is not true for infidelity. A woman is much more likely to have penetrative sex with a man other than her partner during her fertile phase. Moreover, she is much less likely to use or insist on the man using contraception on such occasions.

In order for these patterns for infidelity to be detected statistically, women must have cycles of mood and behaviour that promote them. We can see a hint of how this might occur, not only in the scene we have just witnessed, but also in the first scene involving this couple. The first month we met them, the woman did not conceive. She was cool towards her partner during her fertile period and we suggested that her lack of conception was a success for her body, not a failure by the couple. At that time, the only man available to her was her partner, and her body had decided – for the conscious brain is not involved here – that it was not yet a good time to give him the paternity of her first child. To achieve this, it had generated a feeling of coolness towards him for the duration of her fertile period.

This month, a chance meeting with her ex-boyfriend meant

that the woman had an alternative potential father for her first child – an alternative which her body actually preferred. She had two opportunities to be unfaithful, first on the Thursday, then on the following Wednesday. She took one opportunity, but not the other. The Thursday evening had been instigated by her ex-boyfriend and the woman was in an infertile phase of her cycle. Her body found barely any attraction in sex with this man. Her coolness that evening had kept him at arm's length, even though he was manifestly hoping for sex. On the Wednesday, however, her mood was quite different. As she moved into her fertile phase on the Monday, the idea of seeing her ex-boyfriend began to appeal to her. Her motivation to do something about it, however, did not peak until the Wednesday. Unknown to everyone, including herself, this was the day that insemination was most likely to lead to conception.

When the Wednesday evening came, her mood and body language were very different from the previous Thursday. Most of the leading suggestions still came from the man, but whereas the week before she had made it clear she was not interested, this time she readily cooperated. In his bedroom, she cooperated further to ensure that she obtained his sperm. She accepted penetration and ejaculation with the minimum of foreplay, and ignored the question of contraception completely. Later, her conscious mind would rationalise her behaviour in terms of being overwhelmed by the excitement and passion of the moment. In reality it was simply that, two days in advance of ovulation, her body was eager to collect this man's sperm. The reason her body wanted a *second* insemination we shall see later (Scene 25), but once she had collected his sperm she lost interest in staying with him. Then, by far the most important thing was to get back to her partner.

Subconsciously, there were probably two main reasons why her body engineered such a sudden change of mood towards her lover. In part, the reasoning will also have been in her conscious mind, though she will have given herself slightly different explanations. The underlying strategy her body was pursuing was that, no matter who fathered the child, the man best suited to help her raise it was her partner. It was vital to this strategy, therefore, that any infidelity should not be discovered. Her body generated a fear of discovery and a sense of panic, but needed to recruit her conscious mind to work out the finer details of timing and story-telling. The two together seem to work well. Surveys suggest that one-off infidelities such as this are rarely discovered, and even longer-term infidelities have only a fifty–fifty chance of discovery. On this occasion, as we have seen, the woman succeeded in covering her tracks.

There is one further stage in her body's strategy which her conscious mind is most unlikely to fathom. When she gets home, she works very hard to have sex with her partner. Consciously, she will have seen this as helping to avoid detection. If she can get her partner to inseminate her, any tell-tale damp patch on the sheets or any smell of semen will not arouse his suspicions. What her conscious mind will not realise, though, is that having collected sperm from her ex-boyfriend, her body is now very keen also to collect sperm from her partner. Her body has already decided that, on balance, her ex-boyfriend would make a better genetic father than her partner. The one thing it doesn't know is how their ejaculates compare. She wants to have her egg fertilised by her ex-boyfriend only if his ejaculate is also the more fertile and competitive. The way for her to discover this is to pit one ejaculate against the other. In other words, her body wants to promote *sperm warfare* between the two men, and this is probably her only chance ever to do so.

Once a woman's body contains sperm from two or more different men, those sperm compete for the prize of fertilising her egg. But the contest that takes place is not a simple game of chance, nor is it just a race. It is indeed a war – a war between two (or more) armies. And it is this warfare between ejaculates, or the threat of it, that has shaped the sexuality of every man and woman alive today, as well as the sexuality of just about every other animal that has ever existed.

Sperm warfare is more common and more important than most people suppose. A recent study in Britain concluded that 4 per cent of people are conceived via sperm warfare. In other words, one in every twenty-five owe their existence to the fact that their genetic father's sperm out-competed the sperm from one or more other men within the reproductive tract of their mother. If this does not seem very many, it nevertheless means that since about 1900 every single one of us will have had an ancestor who was conceived via sperm warfare. Every one of us, therefore, is the person we are today because one of our recent ancestors produced an ejaculate competitive enough to win a sperm war.

Most often, because a woman only produces one egg at a time, sperm warfare can produce only one winner, as in Scene 6. Occasionally, however, a woman produces two eggs at the same time and gives birth to fraternal twins. Under such circumstances, a different outcome of warfare is possible – a draw. There are several remarkable cases on record – most obvious when the two rivals in sperm warfare are of different races – in which fraternal twins have different fathers.

Let us go back to the moment of conception that we witnessed earlier. Three sperm arrived at the egg simultaneously. They were all from the woman's lover. If we move back down her oviduct to the collection of sperm waiting quietly in the rest area for their turn to swim up the oviduct, we find that nine out of every ten are also from the lover. The woman's partner had lost this game of sperm wars in a big way, even though he had done all he could in preparation. We are back to routine sex.

So far, we have interpreted a man's interest in routine sex simply in terms of his attempt to maintain a steady flow of fertile sperm through a woman's oviduct. Routine sex, however, does more than just top up the partner with fertile sperm: it also prepares for sperm warfare. Moreover, the level of preparation depends on the risk of war. In deciding how many sperm to load and ejaculate during routine sex, the man's body weighs up the chances that his partner might contain sperm from another man. His body does this very simply by registering how much time he has spent with the woman he is about to inseminate since they last had sex. If they have not had sex for over a week, then he registers how much time he has spent with her over the past eight days.

Crude though this strategy might seem, it does work. The less time a man spends with his partner, the greater the chance of her infidelity. If he spends more than 80 per cent of his time with her, there is virtually no chance of her being unfaithful. But if he is with her for as little as 10 per cent of his time, there is over a 10 per cent chance. In other words, as far as the man's body is concerned, if he has spent less time with his partner since they last had sex, there is a greater chance that when he inseminates her her body will already contain sperm from another man. To increase his chances of winning the sperm wars that might follow, he needs to introduce more sperm. And this is just what he does.

The difference in the number of sperm he will ejaculate, depending on the circumstances just outlined, is quite large. In the scene of infidelity we have just witnessed, the woman returned home on the Wednesday night and had sex with her partner. In deciding how many sperm to introduce, his sleepy body will first have registered that it was three days since they last had sex (Sunday). The average top-up for such a gap is about three hundred million sperm. His body next registered that they had been together about 50 per cent of the time since then, giving an average chance of infidelity. So the average top-up of about three hundred million is about right, and that's the number of sperm his body should have loaded and then ejaculated. If the couple had been in each other's company continuously during those three days, the risk of her infidelity would have been near zero and he would have introduced only about a hundred million sperm. On the other hand, if one or other of them had been away from early Monday morning to late Wednesday night, the risk of her infidelity would have been greater and he would have injected about five hundred million.

For the woman's lover, the situation was quite different. It was the first time he had inseminated this woman for over six years. Moreover, even over the last eight days he had spent only a few hours with her. His body judged, correctly of course, that her body had a very high chance of containing sperm from another man, and will have responded by loading and ejaculating six hundred million. Half an hour later, he injected a further hundred million or so. As far as 'Infidelity Wednesday' is concerned, sperm wars began with the lover putting a sperm army into the woman's tract that was twice the size of her partner's.

So the lover had a twofold advantage over the partner from the very beginning. By the time the war was coming to its climax and the prize of fertilisation was nearly ready to be claimed, however, the lover had a ninefold advantage. (Remember that in the rest area of the woman's oviduct, nine out of every ten sperm waiting to set off for the fertilisation zone were the lover's.) What had happened during sperm warfare to shift the odds even further in the lover's favour? The first step in finding out is to meet and appreciate the soldiers themselves – the sperm.

The most common sperm in the human ejaculate is, of course, the magnificent, sleek athletic cell that most people know, with a head, a mid-piece and a long, slender tail. The head is paddle-shaped, oval in outline but flattened and wearing a cap. This cap is filled with important fluids. Inside the head, densely folded, is the package of DNA, the genes that a fertile sperm will deliver to the heart of the egg. The head perches like a lollipop on the short, stiff mid-piece which is the sperm's powerhouse, the place where stored energy is mobilised to activate the tail for swimming. These sleek individuals travel effortlessly through the female's body fluids, pushed forward by elegant waves whipping in slow motion down the length of their tail.

Familiar though this image will be to most people, such sperm comprise only just over half of the total in a normal ejaculate. A sperm army is a much more motley collection of characters than most people expect. For example, some sperm have a big head, others a small one. Yet others, the pin-heads, have a head that is so small that there is no room to carry the genetic package of DNA. Some sperm have a round head, some a cigar-shaped one, some a pear-shaped one, some a dumb-bell shaped one, and some a head that is so irregular in shape that it defies description. Some sperm, the real monster troops, have two, three or, very occasionally, four heads.

The shape of the head is not the only feature to differ. Some sperm have short tails, some have tails coiled like a spring, and some have two, three or occasionally four tails. Some sperm, the hunchbacks of the force, have mid-pieces that are bent into a right-angle. Yet others, like hikers carrying a rucksack, have bags of cell material on their mid-pieces. On average, only about 60 per cent of the army are the familiar, sleek athletes; the remainder are this collection of deviants. All, however, have an important part to play in sperm warfare.

On 'Infidelity Wednesday', it was the performance of the lover's army as a whole that helped to increase the odds in his favour, as the sperm war progressed, from 2:1 to 9:1. To find out how, we need to climb into the woman's body with a microscope and follow the war in detail through all its campaigns. We will begin at the moment the woman and her lover walk into his hotel bedroom, undress and begin to have sex on the floor.

SCENE 7

A Sperm War

As the woman and her lover sink to the floor, only moments away from intercourse, her body already contains sperm. Her partner inseminated a total of six hundred million during their routine intercourses the previous weekend. Most were ejected in her various flowbacks, but even so some are still inside her. Their ability to influence the outcome of sperm warfare, however, depends on where they are.

A few ineffectual sperm are at the top of her vagina, carried there by cervical mucus, which has been dripping out of her cervix and oozing down her vagina all day in anticipation of this moment of infidelity. Each drip of mucus carried with it a few of her partner's sperm. As these sperm were lost from the future battleground in her cervix, they were partly replaced higher up by the last handfuls of sperm from her cervical crypts. These emerged and entered her mucus channels in a vain attempt to make good the numbers being lost into her vagina. However, the numbers being lost were greater than the number of their replacements, and all day her partner's cervical defences had been slowly declining.

The sperm lodged in her cervical mucus are not the sleek types already referred to. Instead, they are sluggardly *blockers* – sperm whose role is to prevent any later sperm from passing through to her cervical crypts and womb. Sperm with coiled tails, a bent mid-piece, a large 'rucksack', a large head, or with two, three or four heads can block very effectively any of the very narrow mucus channels in which they lodge. So, too, can two sperm side by side. As her lover thrusts inside her, however, relatively few of her mucus channels are still blocked by this rapidly dwindling collection of her partner's sperm.

These blocking sperm are not her partner's only defence inside her body. Roaming around in the void of her womb are a few more of his sperm, though these too are dwindling in numbers. These sperm look familiar. They are svelte and athletic, but they are not there to fertilise. These are *killer sperm*, roaming around in search of sperm from other men to destroy. Each time a killer encounters another sperm, it tests the chemicals on the surface of the other's head. If those chemicals are the same as on its own head, the killer recognises an ally and moves on to continue its search. So far, all encounters inside this woman have been with allies and the killers' deadly services have not been needed. Many are now beginning to move slowly, and large numbers are dying of old age. The weakest have been in her womb for three days. The more active are more recent arrivals from the reservoirs in her cervical crypts.

The woman's womb is not the only territory stalked by killers. A few more are scattered along her oviducts. There is even one swimming solitarily in her body cavity near to her left ovary. These killer sperm in her oviducts accompany the last handful of the partner's fertile sperm, the *egg-getters*. Killer sperm and egg-getters look very similar. They are both sleek and athletic in form but, whereas the killers have average-sized heads, the heads of the egg-getters are slightly larger. If the woman ovulated now, her partner would still have a good chance of fertilisation. But ovulation is still two days away, and war is about to begin.

After very few thrusts, the lover deposits his seminal pool in the woman's vagina. Her cervix dips into and stays in the pool, and the vanguard of his army begins to stream into the channels of her cervical mucus. This army contains about five hundred million killer sperm, about one million egg-getters, and about a hundred million blockers. Some are denied passage through mucus channels by the partner's blocking sperm. So few of these blockers remain, though, that almost all of the cervical channels are now clear. The invaders pour through in waves. A few hundred egg-getters with support from killers travel straight through the cervix into the womb, heading immediately for the rest area in the oviduct. The remainder of the egg-getters, some of the killers and the youngest of the blockers – a few million in all – head for the cervical crypts. They pour in, settle down, and await developments. The remainder of the killers travel more slowly through the cervix into the womb, leaving behind the much slower blockers. These latter distribute themselves throughout the mucus channels, then settle down, many immediately coiling their tails as if in anticipation of a long wait.

Some of the vanguard of the lover's egg-getters do not make it to the oviduct. As we have seen, the partner has relatively few killer sperm still active in the womb, but those that are there do their best to stem the lover's tide. As soon as a killer from either man first encounters a sperm from the rival, it is alerted that war has begun. For an hour or so, the killers from both men swim much faster than normally, seeking out as many rival sperm as possible. Their aim is to poison the rival's egg-getters and killers using the deadly cocktail of fluids in the cap which they each carry on their heads. They do this via head-to-head combat. First, as we have seen, they probe with the tips of their heads at every sperm they encounter, comparing its surface chemicals with their own and checking for similarities and differences. If a killer finds a sperm from the rival's army, it tries to jab the deadly *tip* of its head against the vulnerable *side* of its opponent's head, applying a small amount of corrosive poison with each jab. Having jabbed several times it moves on, leaving the other sperm to die.

A single killer sperm carries enough poison to kill many sperm from the rival army, but gradually its cap runs out of chemicals, for it has no reserves of energy to make any more. In a last-ditch attempt to kill just one more sperm, it tries to stick its head to a rival's and apply the last drops of its lethal fluids. As the war progresses, there is a gradual increase in these pairs of dead and dying sperm, joined at the head in a deadly embrace.

In this initial skirmish, one or two of the partner's killers do their job and some of the lover's egg-getters and killers die from head-to-head combat, their heads coated in the poison. Any initial success of the partner's sperm, though, is short-lived. In their turn, they are found by the invading bands of killers from the lover which accompany his egg-getters. In a frenzy of kamikaze mayhem, killer sperm from both sides attempt to annihilate each other's troops. Outnumbered by at least a thousand to one, however, the last remaining sperm from the partner are soon killed.

The battlefront now moves into the oviducts, where the killing continues. With only a few losses of their own, the lover's sperm systematically wipe out the partner's last remaining egg-getters and killers. By the time the woman and her lover have sex again, an hour later, the first battle is over and none of the partner's sperm remain alive inside her. Actually, the role of the second insemination in this particular sperm war is more complicated than it seems – but here that discussion would be a distraction.

So far, our war has been so one-sided that it has been little more than a brushing aside. The main battle is still to come. It begins when the woman gets home and her body urges her to sit astride her partner, push his penis into her vagina, and stimulate him to ejaculate. By doing so, she has staged a *real* war. Nevertheless, even though her partner now enters a fresh army of three hundred million sperm into the arena, the contest is still going to be one-sided.

As soon as the newly introduced sperm from the partner attempt to leave the seminal pool, they encounter problems. The channels of the woman's cervical mucus are nearly all blocked, not only with sperm from the lover, but also with white blood cells from the woman herself. The huge number of the lover's sperm and the matching number of white blood cells are doing their job almost perfectly, and the partner's sperm are much more hindered in leaving the seminal pool than were the lover's, a couple of hours earlier. Queues of the partner's sperm develop in the blocked channels, producing tailbacks all the way to the seminal pool. As a result, only a small proportion of the partner's army manages to escape the pool before the woman ejects her flowback.

Even those sperm which do escape the pool and find a clear channel are not yet out of trouble. The small vanguard of egg-getters and killers which head straight into the womb find themselves having to run the gauntlet of hordes of the lover's killers. Inevitably, one or two get through without being poisoned, but only to run into further problems as they then try to leave the womb. The entrance into each oviduct is narrow, only just large enough to allow easy passage to a descending egg. Moreover, both oviduct entrances are blocked by lover's sperm and patrolled by killers, and many of the partner's sperm are killed as they try to push through. Even those few that do escape and eventually arrive safely in the oviduct rest area are still at risk to the lover's killers, which patrol the whole area.

Lower down in the woman's tract, in her cervix, many of her partner's sperm are trying to enter her cervical crypts. But the crypt entrances too are patrolled by killers and, in any case, inside they are virtually full of the lover's sperm. Just occasionally, some of the partner's sperm stumble across a channel leading to an empty crypt, but the majority get stranded in the mucus where they fall prey to the combined forces of the lover's sperm and the woman's white blood cells.

Clearly, the Wednesday encounter in this war has gone very much in favour of the lover, and nothing much happens over the next two days to redress the balance. As the woman goes about as normal on Thursday and Friday, the population of blocking sperm from both men in her cervical mucus slowly declines. Some drip into her vagina, carried by mucus. Others are mopped up by rearguard white blood cells. Even when some of these blockers are replaced higher up in her cervix with sperm from the cervical crypts, the new recruits fail to make good the loss and the blockers dwindle. The killer sperm in her womb, after an initial decline from battle losses, actually build up in numbers on Thursday with a fresh recruitment from the cervical crypts. Then the killers also begin to decline in number. A steady flow of egg-getters from both men (but mainly the lover) leave the crypts, heading for the rest areas in the oviducts. En route, they have to run the gauntlet of rival killer sperm in the womb, and most of the partner's egg-getters fail because these killers are almost all from the lover. By Friday night, with only hours now to go to ovulation, the partner's egg-getters in the oviducts are outnumbered by about one hundred to one.

When the woman and her partner have sex on the Friday night, there is only one more hour to ovulation. Now, the partner's sperm have a much easier passage through the cervical mucus because the number of blockers is much reduced. Although most of his sperm head for the now half-empty cervical crypts, a vanguard of egg-getters and killers head straight for the oviducts. Many are killed or slowed down in the womb by the lover's patrolling killers, but enough get through to reduce the odds in the oviducts from a hundred to one to about ten to one. It is at this point that the woman produces an egg from one of her ovaries and a chemical signal passes down the adjacent oviduct. This signal activates hundreds of the sperm in the rest area, and a wave of sperm begin to make their way up the oviduct towards the fertilisation zone. It is now a race, or rather an obstacle race, because there are still killers in the oviduct, mainly from the lover. The partner's sperm, particularly those few which have just arrived straight from the insemination, are actually faster than the lover's – if all else were equal, the partner could still win the prize of fertilisation.

But all else is not equal. One after another, the partner's egg-getters run into the lover's killer sperm. As the egg reaches the zone of fertilisation in the oviduct and the first sperm arrive, the odds against the partner have reduced to five to one, but that is not enough. The first three sperm to arrive are all from the lover, and one of these claims the prize. An hour later, with the partner's fresh sperm now overwhelming the lover's at all points in the woman's tract, the odds in the oviducts swing heavily in the partner's favour. But it is too late. The lover has made it and the woman's daughter, to be born in nine months' time, will not have been sired by the man she will call her father. But nobody will ever know.

4

Counting the Cost

SCENE 8

Doesn't He Look Like His Father?

As the man drifted back into consciousness, he painfully turned over his left hand which had been resting, palm down, on the bed. His partner reached out and placed her hand in his, distressed by the coolness of his skin. As their eyes met, she shook her head, answering his mute question.

The man knew that death wasn't far away, but he couldn't die yet. He had set himself one last task and had to live just that little bit longer. Despite the drugs and the pain, panic rose within him that he might fail and die too soon. His son was on his way, flying from the other side of the world, and the man desperately wanted to see him just one more time. Nothing else would make his last moments peaceful.

His eyes closed and he drifted off once again into semi-consciousness. Scenes from the past opened and closed so vividly he could swear he was actually there. He walked into the room where he met his lifelong partner and saw her for the very first time. He saw the blood and water as his son shot that last short distance out into the world. The midwife picked the baby up, identified his boyhood, and in the next breath remarked how much he looked like his father. Then he was wrapped and placed in his father's arms, his tiny, wizened face pointing upwards, bottom lip quivering as he sucked at a non-existent nipple. It was the most emotional moment of the man's life, his own flesh and blood there in his arms.

The man opened his eyes again. Still all he could see was his partner. After their son had been born, she had never really wanted any more children, but he hadn't minded. Just having the one child meant that they had never needed to stint on their son's comfort, development and education. At the same time, they had found it relatively easy to become modestly wealthy. Their investment of time and money had been more than repaid by their son's successes.

Three times, as his son was growing up, the man was almost tempted into infidelity. But each time, at the last moment, he had resisted for fear it would break up his home. He would have been sad to lose his partner, but he would have been heartbroken to lose his son. The two of them had always been close. They had shared all of those things that a father and son can share, even through the boy's difficult adolescence. He relived the pride he had felt at seeing him graduate, then watched once more as his career went from strength to strength. He met again the succession of pretty girls who clamoured for his attention, and the beauty who was to become like a daughter to him. He remembered the surges of grand-paternalism as, one by one, they had given him five grandchildren.

As real as if it were actually happening, he felt himself lift the photograph which was now by his hospital bedside but which for years had had pride of place in his lounge. After his son's emigration, prompted by a career move apparently too lucrative to refuse, the picture had taken on a special significance. It was of his dynasty, as he called it; a professional photograph of himself, his son and the five grandchildren. As he never tired of saying, the picture showed his contribution to the world and to future generations, a contribution more lasting than any work of art. His son and his grandchildren had already inherited his genes. Now, very soon, they would inherit a large part of his wealth.

His heart skipped a beat as he thought he saw a young man come into the room. He was sure it was his son, and he looked so well, so successful – and so strangely young. The man smiled. He had done it. He had hung on just long enough.

His partner knew he was dead. She had felt his hand growing colder and colder. Now he was gone. She thought she had used up all her tears, but more came. After a while, she called for a nurse; then, after a few more moments of contemplation, left the room to wait for her son. He finally arrived two hours later. After she had broken the news to him, the pair of them went in and stood over the man's body, now totally cold. The woman tried to console her son by telling him that in the man's last hours, during his few moments of consciousness, he had spoken of nothing else but of him and his family.

While openly weeping, her son cursed the airport delays and heavy traffic that had made him too late. Then, in an outburst he would later regret, he turned on his mother and swore at her. He cursed her infidelity and lamented the day that she had saddled him with her secret. For ten long years she had made him keep up the pretence until, in the end, the burden had become too much and he had felt driven to emigrate. But most of all, he cursed her for making him hate himself today. During the long flight home, a single thought had plagued his mind. Why am I bothering? – he's not even my real father.

~

On many occasions in this book, we encounter people who enhance their reproductive success via infidelity. Such behaviour, however, is only advantageous if the person concerned manages to gain the benefits of infidelity without incurring its even greater costs.

This chapter is concerned with the potential costs of being

unfaithful. It consists of Scenes 8 to 11, each of which explores one or more of the costs and dangers of infidelity. Some of these dangers are experienced by the person who is being unfaithful; some by their partner. In this first scene, we examine the reproductive repercussions for a man of being unknowingly tricked into raising another man's child.

The comments that people make when they are first confronted by a new-born baby show a surprising preoccupation with seeking a resemblance between the baby and the presumed father. It is not known how often such comments and comparisons are accurate. In Scene 8 the midwife was wrong, but the man would have found it reassuring none the less. But, as life turned out, it might have been better if he hadn't – he might have been more likely to retrieve his situation, reproductively.

In his generation's cruel competition to pass on its genes, the dying man was a reproductive failure. For him there were no descendants; no dynasty. He had been outmanoeuvred in life's mating game by his partner and a man he never even knew – the man who was the real, genetic, father of his 'son'. Between them, the two had tricked him into dedicating all of his reproductive effort into raising a child who wasn't his, just like the small bird that is tricked into raising a monstrous cuckoo chick.

Had he not been duped in this way, there was in principle nothing wrong with his strategy of having just one child. Recent studies have shown that, all else being equal, increasing one's wealth and investing more into each child can increase reproductive success just as much as having more children. It does so because each child then has a greater chance of survival, grows to be healthier and wealthier, and so becomes more likely to attract the opposite sex. Eventually, such a child should produce more grandchildren or great-grandchildren than a child who received less investment from his or her parents.

Sons, in particular, make good investments (Scene 18). Wealthier, healthier sons have more opportunity to inseminate girls before they choose a long-term partner, are more likely to obtain an attractive, fertile and faithful partner, and are more likely to have the opportunity for infidelity. Even apart from the grandchildren such a son might produce through his long-term relationships, he is also more likely to produce 'satellite' grandchildren via other women, often in the process tricking other men into raising his children as if they were their own.

The greatest reproductive success is achieved by people who strike the best balance between the pursuit of wealth and status and the production of children. This principle applies just as much to an African cattle-herder as it does to a Western industrialist. It also applies to other animals. A male bird, for instance, has to strike a balance between gaining a better territory and feeding its young on the one hand, and finding opportunities to mate on the other. The best balance, of course, can be elusive. Spend so long accumulating resources to invest that you never actually find time to reproduce, and your strategy will fail. Spend all your time having children and none accumulating resources, and your strategy again fails. Your children may die of malnutrition or become so unhealthy and disease-ridden that they become unattractive or infertile.

The single-child strategy, which is the ultimate in investment, can be successful as we have seen – but it can also fail. Moreover, when it does fail, it does so spectacularly. If that child dies through accident or disease or is infertile through some misfortune of genetics or infection, the single-child strategy is a total failure. Or, if your situation is like that of the man in Scene 8, the single-child strategy is again a total failure.

For the woman in the scene, however, the strategy worked wonderfully. She produced a son who survived and avoided major diseases. Moreover, through his receipt and use of the higher-than-average family wealth, he was able to achieve a status in terms of health and wealth that made him a popular target for attractive and fertile young girls (Scene 18). For all his mother knew, her son might have produced children with some of these other women. He might even have tricked other men in the same way that his genetic father had tricked the man who had just died. Even apart from such potential satellite children, her son had successfully produced five children with his long-term partner. Had the woman had more than the one child, the reduced investment in each could have led to her having fewer grand-children. As events turned out, her strategy was a good one.

The strategy also worked well for the genetic father. Not only did he enjoy the same reproductive benefits through his son as did the woman, but he undoubtedly enjoyed further reproductive success with his own long-term partner. Biologically, his success contrasts with the failure of the man who raised his, the genetic father's, child as if it were his own. Having been deceived early on in his reproductive life, the latter had several opportunities to retrieve the situation but, as events turned out, he responded disadvantageously. He could have been more persuasive in chan-ging his partner's mind about not having more children, but he wasn't. He could have taken the opportunities for infidelity that presented themselves, but he didn't. He could even have left his partner and tried a long-term relationship with another woman who would have given him children, but he didn't. For him, the dangerous strategy of having only one child was a disaster.

Had the child he was raising been his own, his character and his responses to the challenge of child-rearing would have been advantageous. The couple would *both* have reaped the rewards of a successful son. But because the child was not his own, biologically his genetic package of characteristics was a failure. With all the cruelty that is natural selection, his genes were weeded out, never to be passed on to future generations.

The experience of the dying man in Scene 8 is by no means rare. World wide, it has been calculated from studies of blood groups that about 10 per cent of children are in fact not sired by the man who thinks he is their father. This is also the level found in industrial Western societies (see Scene 18 for further details). There is a real need for an extensive study using modern techniques such as DNA fingerprinting. So far, the nearest thing to such a study comes from the paternity tests carried out by child support agencies. They are responding to absent 'fathers' who demand such tests in an attempt to avoid or delay the enforced financial support of an ex-partner. Internationally, child support agencies are reporting a non-paternity rate of about 15 per cent.

All figures for non-paternity are of the proportion of children actually *born*. The non-paternity level for children *conceived* will be even higher. This is because a woman is more likely to abort a child conceived via a man other than her long-term partner. Almost certainly this happens primarily when her partner either knows or has a good chance of finding out that he is not the real father. The abortion then represents an attempt by the woman to avoid the costs of infidelity discussed in Scenes 9 and 11.

Although properly controlled DNA fingerprinting studies have not been carried out on humans, they have been carried out on a wide range of apparently monogamous birds. The results suggest a roughly 30 per cent incidence of males raising other males' offspring, comparable with but slightly higher than the level in humans. So it would seem that the average male bird has even less reason to be reassured by a passing resemblance to its offspring than has the average man.

SCENE 9

Making Mistakes

The woman turned the corner, then stopped under a street lamp to check that her keys were in her bag. A few minutes more would see her home, and her spirits were sinking fast. Although the night was cold, she paused for a while under the light – anything to delay the moment of arrival.

She had spent the evening at her sister's, about fifteen minutes' walk away, seeking peace and asking advice. Her sister was in no doubt.

'Take the children and go,' she had said. 'Stay with mother – she'll have you while you sort things out.'

Still standing under the light, the woman fingered her cheek bone. The tenderness had nearly gone, but she knew it wouldn't be long before the bruises were back again. She took a deep breath, braced herself, and walked the remaining distance to her front gate. She had hoped he would be in bed by now, but as she went up the path she could see that the sitting-room light was still on.

Her partner didn't look up when she walked into the room, his gaze staying firmly fixed on the TV screen. There was a can of beer in his hand, and a further eight crushed empty ones littering the floor. She recognised the atmosphere only too well, and knew she had to be careful. For a while she busied herself with minutiae, tidying up the debris of his evening at home with their two children. In the end, she could stand the silence no longer and asked, as calmly as she could manage, if the children had gone to bed without fuss. Without looking at her, he spat his reply. *Her* children had gone to bed fine. *He* didn't have any.

She knew better than to contradict him. Both of the children were in fact his, but recently he had decided they were not. Since then, he hadn't missed an opportunity to voice his new-found doubts – to her, to the children, to the neighbours, in fact to anybody who would listen. She sighed more aggressively than she intended, then said that if he was going to start that again, she was going to bed.

'Come here,' he ordered.

She hesitated, a familiar fear rising inside her.

'Come here,' he repeated, even more forcefully, still not looking at her.

She knew she had no choice. Running away only made him worse. She walked over and stood in front of him. He remained seated.

'You've been with him again,' he said flatly, staring through her stomach at the television behind.

She said she hadn't been with anybody, only her sister. There wasn't anybody. He should phone her sister if he didn't believe her. He didn't need to phone her sister, he said – he *knew*. She told him he was drunk and crazy, that there was no talking to him any more, and that she was going to bed. Now, for the first time since her return, he looked up at her face and told her to stay where she was. She saw the wild look that always came into his eyes before he attacked her. As he demanded to know who she had been with, fear gripped her. He said he knew she was screwing around. She had always screwed around. He just wanted to know who it was, so that this time he could kill him.

As he staggered to his feet, she backed away, pleading with him not to hurt her again. He demanded even more loudly to know who she was screwing. She half cried, half shouted that there was nobody. Ignoring her, he yelled back that whoever it was he wasn't going to have a free ride. He told her to take off her clothes and accept what was coming to her. She whimpered 'No', but it was no good. His fists and hands were everywhere as he manhandled her to the floor. As new bruises were added to old, the pain was familiar but more intense – so intense that it was almost a relief when he stopped beating her to tear off her underwear. Then he unzipped his trousers and entered her, thrusting vigorously and painfully.

Almost as soon as he had inseminated her he stood up, telling her that if he ever found out who she was seeing he was a dead man. If she was lucky, he might let *her* live. Now, he was going to bed and she could sleep on the floor where she belonged.

The next day she packed her and her children's belongings and walked out, taking her sister's advice and moving in with their mother. Over the next few weeks, she considered legal action for violence and rape but eventually decided that her best option was to have nothing further to do with him. She never saw him again, and he never tried to contact her or the children. Later, she heard that only two weeks after she had moved out, a girl of about nineteen had moved in. A year later, he was dead, murdered by the nineteen-year-old's father.

The woman stayed with her mother far longer than either of them wanted. Eventually, she moved in with the first man who was prepared to take on her and her children. Having been deserted by his previous partner after a barren relationship, he had no children himself. At first, he was a good father to hers. But no sooner had she given birth to his daughter than his attitude changed. He began to hit her son and, unknown to her, to sexually abuse her elder daughter, who became morose and rebellious. After a particularly savage beating from his stepfather, her son left home to live with his grandmother until he could support himself. Many years later he was diagnosed as infertile, the result, he always maintained, of that final beating. Her elder daughter, once in her early teens, simply ran away and never again re-established contact.

The woman never forgave her new partner for driving away her older children, but rather than repeat the mistakes of the past she stayed with him. Once her son and elder daughter had left, her relationship with him seemed to improve. They had another child and managed to raise their family without too many problems.

～

There are pros and cons to both fidelity and infidelity. From the viewpoint of reproductive success, neither is inevitably advantageous or disadvantageous. What is always disadvantageous is making mistakes. Choosing an unsuitable partner from among those available is one mistake. Another is to misjudge a situation and be unfaithful when it would be better to be faithful – and to be faithful when it would be better to be unfaithful. Yet another is to be under- or over-zealous in attempting to prevent a partner from being unfaithful. The people who do best in life's generation game are those who judge their situation correctly and then respond appropriately.

A man has a lot to lose from his partner's infidelity. First, he may be tricked into devoting a lifetime of wealth and effort into raising another man's child (Scene 8). Secondly, he is at greater risk to sexually transmitted diseases, because his partner is at greater risk. Thirdly, he risks being deserted by his partner if she judges the other man to be a better prospect – in which case, she may either take their children with her or leave them behind. Whichever she does, his reproductive success may suffer.

If she takes his children, they may be raised by a stepfather, which could be reproductively costly. Studies of many different cultures around the world have all shown that children are much more likely to be abused or even killed by a stepfather than by their genetic father. But if his ex-partner tries to raise their children on her own, this too can be costly. In at least some societies, the children in single-parent families have a higher mortality risk. The same risk exists if his partner leaves their children for *him* to raise single-handedly. Even if he manages to attract another partner, he still risks having them abused or murdered – this time by their stepmother.

Humans aren't the only animals known to behave more aggressively towards their stepchildren. Male lions provide a good example. Lion prides contain two or three males and up to eight females and their young. Wandering over the savannah are bachelor groups of two or three males, each group looking for a pride from which they can oust the current males. If they succeed, the first thing they do is kill the cubs, the pride's legacy from the previous males. The females are brought into heat by the loss of their young, giving the new males an early chance to sire their own offspring.

Some monkeys that live in harem groups – one male with several females – behave in the same way. If the harem-owner is driven out, the new male kills the previous male's offspring. Even monkeys that live in much larger groups, consisting of many males and females, show similar behaviour. More often than not, a male acts aggressively towards young which *cannot* be his. On occasion, however, he may do the exact opposite and go out of his way to help a female look after such young. But this is not the act of selflessness it might seem – it is instead a ploy to gain sexual access. Once the female produces a baby that could be the male's, he behaves more aggressively towards her older off-spring. Note the similarity of such behaviour to that shown in Scene 9 by the woman's second partner.

There are two ways in which a man may try to reduce the chances of his partner's infidelity. Crudely, these masquerade as possessiveness and jealousy. As noted in Scene 6, on the one hand he can spend so much time with his partner that she has very little opportunity to be unfaithful. On the other, he can threaten her with undesirable consequences if she *is* unfaithful. His two main threats are first, desertion, and secondly, violence – towards her, her lover, or both. The man in Scene 9 used violence, and carried it to excess. Such behaviour is not uncommon.

Infidelity and, just as potently, the suspicion of infidelity are two of the main causes of domestic violence. They are given as the reason for over half of reported cases of the abuse and murder of a partner. In this, humans are for once relatively unusual. Studies of other animals, mainly birds, have shown that males rarely respond to a partner's infidelity with aggression, though they may behave aggressively towards the other male. Monkeys and apes are a little more similar to people. Even so, they are still less predictable than humans in showing aggression towards an unfaithful partner.

Why humans should show more aggression than other animals in such circumstances is unclear. One possibility derives from the differences in size and strength between men and women. In most animals, there is little such difference between the sexes. Potential victims can usually give as good as they get. In humans, however, men do have a considerable physical advantage over women, and most domestic violence is perpetrated by men on women, rather than vice versa. This physical difference gives men more potential than the males of other species to be violent to their female partners.

Measured aggression, or its threat, may play an important part in preventing a partner's infidelity. But if it is misjudged, either in timing or in vigour, the reproductive costs may greatly outweigh the reproductive benefits. This is because there are at least three potential disadvantages to behaving aggressively towards one's sexual partner. First, as long as a male (or a female) needs the partner's cooperation in having and raising children, then that partner's continuing health and fertility are essential. Violence, and any damage it may cause the partner, are therefore disadvantageous to the aggressor as well as to the victim. Secondly, there is the danger of such vigorous retaliation on the part of the victim that it is the aggressor who eventually ends up the more damaged. Finally, other individuals with the victim's interests at heart, such as parents, siblings or potential partners, may also offer a vigorous defence, as the violent man in Scene 9 found to his cost.

This man showed a level of aggression towards his partner unusual among animals. His urge to inseminate her on suspecting infidelity, though, is less unusual, and has its counterpart behaviour among many other species. Male birds, for example, on seeing their partner having sex with another male, fly straight at that male and knock him off. They then immediately inseminate the female. Male rats and monkeys respond similarly. After inseminating his partner, a male rat or monkey would normally wait a while before doing so again. However, if he sees her mating with another male, he re-inseminates her immediately. Even if they have simply been apart for longer than usual, he is more likely to inseminate her when they next meet.

Such rapid re-insemination is an important strategy for success in sperm warfare (Scene 21). If the second male waits too long, more of the first male's sperm have time to leave the seminal pool. In addition, the first male's army has the chance to distribute itself throughout the female's tract in an optimum way – blocking cervical channels, filling cervical crypts, and organising the next few days' passage of killers and egg-getters. On the other hand, if the second male can inseminate the female quickly enough, his own army is there in time to compete for prime positions in the female. Studies on rats have shown that virtually every second counts. The longer the second male waits before inseminating the female, the more pups in the litter will be sired by the first male. Speed is everything.

It is for this reason that rats, monkeys – and men – become sexually excited by the sight of another couple having intercourse. The male's penis becomes erect and he may even load his urethra with sperm. Along with almost all other aspects of human sexuality, the appeal of hard-core pornography owes its existence to behaviour shaped for success in sperm warfare.

Men seek to avoid the costs of their partner's infidelity by being vigilant for signs, minimising opportunities, and threatening desertion or retaliation. If signs of infidelity are detected, then guarding and threats are escalated. Only if infidelity actually occurs, however, are the threats usually carried out. Even then, a man may decide that his long-term prospects for reproduction are still better if he stays with his current partner than with anyone else – in which case, the threats are curtailed.

In Scene 9, separation was costly to both the man and the woman. Their son, at least, may have suffered reproductively as a direct consequence of the separation. The daughter, too, may have done so, though her fate is unknown.

Had the violent man actually been correct in claiming that neither child was his and that the woman was being unfaithful, his extreme behaviour might still have been advantageous. Starting again with a new and younger partner could have been a better strategy than staying with an unfaithful older partner. But he was wrong and, as it happened, his misjudged violence reduced his reproductive success in two ways: first, via the two children he *had* sired; secondly, by leading to his early death it robbed him of the chance of producing further children.

Although deserting her first partner was costly to the woman through its effect on her children, she probably made the best of a bad job – in so far as her reproductive survival was concerned, anyway. Had she stayed with him any longer, his violence could have been so physically damaging that she might have been unable to have more children. He might even have turned on her children, acting on his mistaken conclusion that he was not their father – in which case, their prospects would have been no better or worse with their real father than with a stepfather. Given her situation, the woman was always going to be in danger of achieving few or no grandchildren via her first family. However, by leaving her violent partner when she did, she at least managed to produce a second family, this time with every chance of having grandchildren.

SCENE 10

Licking Infidelity

In the first few moments after it was over, they felt awkward. Having picnicked in the sun in a secluded hay-field, they had been carried by sexual excitement to the brink of infidelity. But she had resisted and in the end, with her encouragement, he had simply ejaculated on to the grass. Now, as passion and frustration subsided, so did the awkwardness. She promised there would be other occasions and better moments. Soon they began to relax.

They stayed in the field for another half-hour, him stretched out in the sun, her sitting in the shade and occasionally talking. Reluctantly, they eventually packed up and drove back to their home town. He dropped her in the car park at work, and then they drove home separately to their respective partners.

As soon as the man walked through his front door and exchanged greetings, he sensed a strangeness in his partner's mood. The phone rang, but when he answered the caller hung up. Over dinner, his partner showed no interest in his day. He told her where his business had taken him and who had gone with him, but got no reaction. All evening, she busied herself, spending scarcely any time with him. On the few occasions they spoke, she hardly looked at him. When asked how her day had been, she shrugged and said it had been normal. She had taken the children to and from school and done the shopping; otherwise she had been in all day. The highlight had been an hour's sunbathing early in the afternoon. Then, for no apparent reason, she volunteered the information that the bed had needed changing and she had washed the sheets. At the end of the evening, as they went upstairs, she complained of how hot the day had been and said she felt like a long bath.

While his partner bathed, the man lay naked on their bed, enjoying the humid heat of the summer night. He thought of the other woman, their picnic, his near-infidelity, and her promise of other occasions and better moments. He became aroused. When his partner came into the room he was visibly ready for sex, but she ignored him and busied herself tidying clothes. Usually, she would walk around naked after a late-night bath, but tonight she was wearing her bath robe. She also switched off the light before undressing and getting into bed, but she probably needn't have worried. The marks on her back were nowhere near as obvious as she imagined.

When he began his sexual overtures, she complained of her tiredness. But he was determined not to be denied a second time that day. For a moment she stiffened and held his head as he worked with his lips down from her breasts to her navel. When he reached her pubic hair, she complained again that she really didn't feel like sex. He told her to relax and offered to lick her in just the way she liked. Tonight, she said, she didn't feel like that either. But it was too late. His head was already between her legs.

He nuzzled her pubic hair, which was sweet-smelling and still slightly damp from her bath. So too were her thighs, but when he began to part her vulval lips with his tongue she galvanised into action, moved down the bed, and began to kiss his face. With hardly any further foreplay she took hold of his penis and guided him into position. He thought it was too soon but, inside, she was surprisingly wet. Intercourse was long. She didn't climax, but he did.

On the other side of town, the man's near-mistress from the hay-field was also having problems with her partner. Once their children were in bed the pair had sat on the settee wearing only their dressing-gowns, supposedly relaxing before going to bed. They each had a cat curled up on their lap. The man had been irritable all evening, and she knew why. He was not happy that she had been on a business trip all day with another man. Every so often since they had eaten he had tried to ask about her day, but the children and their evening routine had prevented any sensible conversation. Now, quiet at last, he sought reassurance.

He asked how she had got on with the other man, commenting that a whole day was a long time to be in somebody's company if you didn't know them very well. She replied that it had been OK, said truthfully that they had talked mainly about work on the way down, then lied that she had slept most of the way back. He complained that they should have gone by train, disliking the idea of their having been alone in the privacy of a car. The woman shrugged, saying she had had no choice. The man had wanted to drive, and he was the boss. Then, in an inspired moment, she said she was glad they had driven. It meant they could open the windows. It had been so hot – and the man had smelled really bad all day.

There was a pause, so she pushed home the advantage gained from her lie. She complained that the man had really annoyed her over lunch. He had seemed to disapprove of everything she liked and believed in. At times, it had been really difficult not to tell him what she thought of him. Her partner relaxed a little, somewhat reassured by her lies.

Suddenly, she had an idea. Gently pushing the cat off her lap, she swivelled round towards her partner and placed her hand partly on his thigh and partly on the cat still on his lap. Making up the story as she went along, she told him of a dream she had had while dozing in the car on the way back. As she talked, she stroked the cat on his lap, apparently accidentally rubbing her partner's penis under his dressing-gown with the back of her hand. In graphic detail, she described how, in the dream, the cat now on his lap had been giving her oral sex. Its tongue had been rough, but gentle. She had got really excited and nearly come, feeling embarrassed when she woke in case the driver had noticed her flush. Now with her partner's full attention, she continued that the thought of oral sex had been in her mind ever since. Complaining that it was a long time since he had obliged her, she urged him to do so now.

As usual, she got her own way. She didn't climax from his licking, but pretended that she did. Then, with the charade over, she let him have intercourse.

~

Humans are not, of course, the only animals to indulge in oral sex. Most male mammals, from rats and dogs to elephants and monkeys, nuzzle, smell and lick the female's vulva during foreplay. Monkeys also touch a female's genitals, sometimes inserting their fingers into her vagina, then smelling and licking them on withdrawal. What all these males are doing is collecting information. They are seeking the answer to three questions. Is the female healthy? Is she fertile? And has she recently had sex with another male? A man is doing exactly the same – and the information he collects can be a big help in his pursuit of reproductive success.

If a woman's secretions smell or taste particularly unpleasant, the male may lose interest in penetration altogether. The smell may indicate disease. Such information is most useful when a man is contemplating sex with a new woman. And because diseases can come and go, it is also worth a man testing the health of even a long-term partner's secretions from time to time.

In many mammals, the smell of the female's secretions is clearly different and more pleasant to males during her fertile phase. In women, though, and in other mammals which hide their fertile phase, the change in smell during the cycle is much less apparent. In one experiment in the USA, a group of volunteer women each wore a tampon overnight on different days of their menstrual cycle. The tampons were placed in open tubes, so they could be smelled but not seen, then given to a panel of assessors. The smell of each tampon was ranked on a scale from 'very unpleasant' to 'very pleasant'. Pleasantness of smell changed with stage of the menstrual cycle, being most unpleasant during menstruation. On average, the smell was slightly, but not consistently, more pleasant during the fertile phase. So by nuzzling a woman's vulva, a man can at least tell if she is menstrual.

As long as a woman's secretions pass a man's health and fertility test, the main piece of information he is seeking is whether she has recently had sex with another man. The man's body can then use this information to change the number of killers and egg-getters he introduces into hers. As we have already seen, crude adjustments are made on the basis of what proportion of time he has spent with the woman over the past week or since their last intercourse (Scene 6). The less time he has spent with her, the greater the *possibility* that she might contain sperm from another man and the more sperm he introduces. This method works, but it is crude. If his body could *know* whether sperm from another man are present, it could make a much finer adjustment to its own sperm numbers.

In Scene 10, the man came home to a partner who, at some time during the day, had clearly been unfaithful to him. She had covered her tracks as best she could by changing the bed, washing the sheets and preventing him from seeing her back in case he noticed the small marks left by her lover. Most important of all, however, she had also had a long bath to try to remove all traces of her lover from her pubic hair, thighs and vulval lips. In all of this, she was obeying her conscious mind, albeit driven by a body that wanted neither to be beaten nor deserted.

Her body had ejected her lover's flowback many hours ago, but some traces of semen always remain in the vagina, sometimes for up to a day. At first, she had tried to avoid sexual contact with her partner completely. Her body favoured the lover and wanted to give his sperm as easy a victory as possible. At the same time, though, she did not want to lose her partner's support. Consequently, when he began to nuzzle her vulva, and the risk of her infidelity being discovered increased, she changed her strategy. She switched from avoiding intercourse to precipitating it solely to distract him from oral sex.

On the other side of town, the woman had returned home to a suspicious partner. She had the advantage, however, that her vagina contained no evidence of how close she had come to infidelity. As a result, she had the opportunity to both reassure and mislead her partner, not only verbally (by criticising the man she had been with) but by forcing him to smell and lick her vulva so as to confirm her innocence. By the end of the evening, both his mind and his body should have been reassured.

Together, the two scenes of oral sex we have just witnessed illustrate the interplay of reassurance and subterfuge that is a hallmark of routine sex between couples. Men may not be able to lick their partner's infidelity via oral sex, but they can certainly collect information useful for deciding what to do next. In the short term, this information can help men to prepare for sperm warfare. In the medium term, it can help them to adjust the intensity with which they guard their mate or search for another partner. In the long term, it can help them to assess the desirability of desertion and, if their partner produces a child, to judge the probability of the child having been fathered by someone else. At the same time, through strategic prevention or encouragement of oral sex, a woman can attempt to reassure or dupe her partner with regard to the actual situation.

Of course, people do not always realise that they are indulging in oral sex for the above reasons. Consciously, what a man thinks he is doing when he licks a woman's genitals is stimulating her into becoming lubricated for penetration. Most often, in the absence of infidelity, what a woman thinks she is doing is seeking sexual stimulation. Both of these things are happening, of course, but they are merely the conscious veneer of the behaviour, not its ultimate function. As in so much of sexual behaviour, the mind consciously pursues superficial stimulation at the behest of a body aiming to achieve much more potent ends.

There is no intrinsic reason why licking a woman's genitals should stimulate her sexually any more than stamping on her foot. Nevertheless, one action is a sexual stimulus, the other is not. What tends to happen in the evolution of sexual stimulation is that unequivocal signs of sexual interest become stimulating, while other signs do not. If this were not the case, males and females would be continually responding sexually to the most irrelevant signals.

For the reasons we have already discussed, male ancestors were driven by their bodies to seek the valuable information contained in and around a female's vagina – the inevitable site of insemination. Failure to seek this information *before* intercourse cost the male in four ways: it meant a greater risk of disease, more chance of inseminating an infertile female, less chance of winning sperm wars, and a reduced chance of combating infidelity. From the female's perspective, these pressures on the male meant that, when a male nuzzled her vulva, it unequivocally signalled his sexual interest. If she was *un*interested, she could simply walk away. If interested, she needed to prepare for intercourse by lubricating her vagina, and so forth. Such advance preparation would make the process of insemination itself more efficient and less damaging. All in all, females who responded to being nuzzled by an attractive male by becoming aroused will have reproduced more successfully than those who did not.

The same arguments apply to the male. If female ancestors allowed him to nuzzle their vulva, instead of walking away, this indicated that they were interested in receiving his sperm. Males who responded by becoming aroused will have missed fewer opportunities for intercourse than males who did not. All in all, males who gained sexual stimulation from nuzzling the vulva of a female will have reproduced more than those who did not.

All of these responses were shaped long before there were humans. When our species first evolved, the males simply inherited from our primate ancestors a predisposition for nuzzling, smelling, licking and fingering the female's genitals. We also, male and female alike, inherited a predisposition to find such behaviour stimulating.

SCENE 11

Checkmate

The bright-red display on the bedside clock changed from 11:59 to 12:00. It was midnight. While his lover had slept her post-coital sleep, the man had been clock-watching. Reluctant though he was to swap the warmth of her bed for the chill air of her room, it really was time to go.

He gently began to wake her by kissing the middle of her back. Their affair had begun half a year earlier in a dark summer wood when, in an unplanned, abandoned moment, they had stripped each other naked and she had helped him to ejaculate. Two months later, soon after she had moved into a bed-sitter of her own, they had full intercourse for the first time. Although they worked in the same building, they rarely met during the day, taking great care to hide their liaison from everybody. Two or three times a week, he visited her bed-sitter for sex. They always went straight to bed. Time was short – they rarely had more than an hour together. Even so, they often had sex twice in a session. He was inseminating her about five times a week, while still inseminating his partner once or twice a week. His lover was eleven years younger than him, but he had learned more about sex in six months of snatched moments with her than he had in ten years with his partner.

For his partner's benefit, he had created a fictitious web of weekly events at work. These gave him routine, almost infallible, alibis which freed him to call in on his lover sometimes on his way to work, sometimes on his way back. Tonight, however, had been special. This freezing late-January night was his lover's birthday and she had insisted on more of his time than the usual hour. Otherwise, she would spend the evening with friends. In the end he had invented a story which, he had to admit, seemed rather good. He had told his partner that a long-term visitor from the other side of the world, on discovering his interest in chess, had challenged him to a match and had offered to cook him an authentic ethnic meal into the bargain. He had reminded his partner that chess took a long time, that they would probably play the best of three, and that it could be a long evening, maybe even midnight before he came home.

The evening had been perfect, a memory for a lifetime. He had cooked his lover a meal and by seven they were in bed, drinking wine, listening to music, and playing with each other's bodies as passion came and went. Altogether it came and went, if with increasing difficulty, four times. The last time he had only pretended to ejaculate, but she had climaxed for the second time that evening. Now, though, their party was over, and ahead of him were two prospects he could do without – running the gauntlet of arrest for drunk-driving and, even worse, the guilt and discomfort of having to fabricate the evening's events to satisfy his partner's curiosity.

He had never quite been able to fathom why he worried so much about being discovered. It wasn't really the prospect of his partner leaving him, though he would miss their two children. In truth, it would worry him more if he lost his lover. Maybe it was his partner's threat to spend every penny he had rather than let him squander it on another woman. Or maybe it was her oft-repeated 'joke' that if he were ever unfaithful she would cut his balls off. He was sure she didn't mean it. But would he ever feel safe again when she picked up a pair of scissors or a knife? Whatever the reason, the prospect of being discovered terrified him.

When his young lover finally surfaced from her sleep, she asked him not to go, gently at first, then with mounting irritation. Finally, indicating more than a little disfavour, she told him he *must* go. He dressed in the cold of her room, kissed her, hesitated a little in a vain attempt to pacify her, then left to walk to his car. He always parked two streets away. The police did not stop him as he drove home, despite a bad few minutes when he thought he was being tailed. Thirty minutes after leaving his lover, just after one o'clock in the morning, he was home – and his partner was waiting for him.

Hardly had he closed the front door than he was being asked almost hysterically what was going on. Taken aback, he apologised for being so late. The games had lasted longer than expected. But he had won in the end, he added, seeking conviction in detail. She paused, then asked him why, if he had been playing chess with his foreign friend all evening, had that very same person phoned a couple of hours ago and asked to talk to him. The man felt his mouth freeze open. Panic swept through him. He couldn't believe his bad luck. In the two years that his colleague had been at his institution, not once had he phoned. How was it that, of all the nights to break his silence, he had chosen tonight?

The man's mind raced as his face reddened and his underarms prickled with sweat. For a moment, he almost told her everything. Then he lied that it had been his assistant's birthday. He had been invited to a men's night out of drink, food, strippers and pool. He knew how much she hated him going on such nights and worried about him drinking and driving, so he had thought it best not to tell her.

She stared into his eyes, trying to work out if he was telling the truth, then went across to the telephone. Pausing, phone in hand, she asked him for his assistant's number. He said he didn't know. In any case, the others wouldn't be back yet, because they had gone on to another club. He had left them to come home because he knew she would be worried if he was too late. She stared at him again. Not knowing how to respond, he hid behind laughter. He said this was silly. He was sorry he had lied to her, but he really had been trying to save her worry. It wouldn't happen again. He would tell her next time – and couldn't they please go to bed because he was drunk and tired.

She moved towards him until their faces were almost touching and stared into his eyes. She told him he was lying – that she knew he was lying. Why didn't he smell of smoke and beer? She paused, then accused him of infidelity. He denied it but she shouted him down, asking who it was. He proclaimed his innocence yet again and objected to her refusal to believe him.

She went quiet for a second, thinking hard. Then she told him to take off his trousers. He protested. She insisted. He said she was being stupid. Angrily, she began to unzip his trousers. He pushed her away, told her to stop, and said that he would do it. As he dropped his trousers and underwear, he hoped against hope that there would be nothing for her to see or smell. But there was. She swore at him, slapped his face, told him to leave her, their children and their house, and said that she never wanted to see him again. When he hesitated, she pushed him, trousers still round his ankles, towards the door.

At first, he didn't resist but at the door he refused to leave. Eventually, she agreed to his spending the night on the settee as long as he had gone by the morning. He spent an uneasy night, unable to sleep partly through discomfort, partly through worry about the future, but most of all because he was afraid. The thought that his partner was about to enter the room, knife or other weapon in hand and intent on revenge, plagued him all night.

He left early, before she came downstairs, but returned home at mid-day. With the children at school, they had a long argument. His partner told him she could never trust him again and that she couldn't live with someone she couldn't trust. He maintained that the evening was a one-off. He said the woman was an ex-girlfriend who had emigrated years ago and then, just last week, had contacted him out of the blue. About to visit the country again for a few days, she wanted to look up old friends and to see him again. He had agreed, not expecting anything to happen after all this time. But they had got drunk, she had invited him back to her hotel and things had just happened.

He was really sorry. He wished he'd resisted, but she had gone now and it would never happen again. He'd had time to think and the fear of losing her and their children had made him realise that all he really wanted was her, his home and their family. He would leave if she wanted but did she realise they couldn't afford to keep the house *and* pay for a flat for him? If she wanted him gone, they would have to sell the house and buy something smaller for her and the children. Was it really worth putting the children through a separation? All because of a night's weakness on his part that would never happen again?

In the end she agreed that he could stay but said he would have to sleep on the bedroom floor. Nor should he imagine he would ever have sex with *her* again. Or that *she* would stay faithful. As soon as she could find another man, she would be gone, taking the children with her.

In fact, he was allowed back in her bed after only a few days and they had sex again after six weeks. He never did tell his lover what had happened and their affair continued undiminished for another six months. She then met a man of her own age and ended their relationship. He stayed with his partner for another five years but during that time, without her knowing, was unfaithful with three women. The first, another girl eleven years his junior, conceived during their affair. But then, after two months, the man she was living with deserted her, doubting that the child was his, and she miscarried. Just briefly, he considered leaving his partner for this younger woman but after some hesitation the girl herself decided she did not want him.

Ironically, despite his four affairs, it was not *his* infidelity but his partner's that eventually ended their fifteen-year-long relationship. As good as her word, as soon as she met another man she left, taking the children with her. She didn't know it but her timing was perfect. His final infidelity had given him gonorrhoea and she departed just in time to avoid being infected herself.

This was the beginning of the worst phase of his life. With treatment, he recovered from his infection but was warned by his doctor, correctly as it turned out, that he might no longer be fertile. He was forced to sell the house so that his ex-partner could take her share of the equity. The cost of supporting his two children combined with losses from two silly financial gambles reduced him almost to penury. His days of affairs with younger girls apparently over, he moved in with a widow, ten years older than himself, whose children were living abroad. After this, life slowly began to improve.

At first his own teenaged children had lived amicably with their mother and her new partner. But when she began a second family, relationships in the household became strained. Eventually, his children moved back to live with him and his new partner. Luckily, she proved to be a good stepmother, treating his children in the few years before they left home as if they were her own.

~

Why is it so important to a woman to prevent or, failing that, detect her partner's infidelity? What are the repercussions of failure for her pursuit of reproductive success?

A woman has a lot to lose from her partner's infidelity. And many, though not all, of the dangers are the same as for a man. First, there is the risk that she will have to share his wealth, time, energy and other resources with the other woman. Secondly, there is the danger that he will eventually leave her for the other woman, reducing his support for her still further. Whichever of them looks after the children, they face the dangers of single- or step-parenthood which we discussed in relation to Scene 9. Thirdly, there is the greater risk of contracting sexually transmitted diseases, because her partner is at greater risk. There is one risk, however, that does not affect her. Unlike a man, she is in no danger of being tricked into raising any children by her partner's lover. This means that, on balance, infidelity is slightly less of a threat to a woman than it is to a man.

This conclusion is as applicable to monogamous birds, monkeys and apes as it is to humans. The result is that, although both males and females of all these species try to prevent their partner from being unfaithful, males are significantly more possessive of their partner than are females. They are significantly more aggressive, and significantly less forgiving. Females may attempt to drive away other females. They may attempt to intervene between their mate and another female. They may even desert their young if they receive too little time and help from their partner. However, since their physical commitment to preventing infidelity is less than the male's, they are less likely to desert if their partner is unfaithful. The same is true for humans.

Women do sometimes wound or kill a partner if he is unfaithful. They do sometimes wound or kill his lover, leave their partner or drive him out, and they also sometimes abandon their children to him to raise. But again, they are less likely to do all of these things than are men and are more likely to overlook discovered indiscretions. Nevertheless, the threat of retribution is always there, and real enough for most men to attempt to hide their infidelity.

Arguably, male animals are never quite as subtle over their

infidelity as females, maybe because the costs of discovery are less. My favourite description of sophisticated infidelity on the part of the female concerns a small brown bird, the dunnock. The male and female, the epitome of contented monogamy, were first seen as they hopped side by side across a lawn, pecking up morsels of food. When they reached a bush, the male went round one way, the female the other. As soon as the bush shielded her from her partner, the female flew in an instant into dense vege- tation nearby. There, she copulated with a lurking male, then flew back to her position behind the bush. A few seconds later, male and female hopped back into each other's view, past the bush. Still intently pecking at morsels, the female acted as if nothing had happened.

Nearly as appealing is the film, shown worldwide, of a female monkey foraging on the ground for food while being watched attentively from a high branch by her consort. Alongside her comes another male. He sits down, innocently picking at himself, hiding his erection from her consort. Every time the consort's attention is distracted, the other male taps the female on the shoulder. In an instant she stands and presents and the male inseminates her. So quick is their intercourse that by the time her consort looks back in their direction, they have resumed their previous activities – innocence personified.

Maybe a female dunnock who flies into vegetation is no more subtle than a male who hides there, waiting. Maybe a female monkey who feigns innocence until her partner is distracted is no more sophisticated than a male who hides his erection and inseminates at the speed of light. And maybe a woman who has sex with the window-cleaner while her partner is at the shops is no more sophisticated than a man who has sex with his secretary in a cupboard at the office party. But, on balance, one always has the feeling that the female of the species, whatever the species, has that slight edge when it comes to ingenuity and imagination. As far as humans are concerned, the best we can use as evidence is that, taking thousands of instances into account, the man is more likely, on average, to have his infidelity discovered than the woman.

In Scene 11, the woman needed to check her mate's genitals only superficially to confirm his infidelity. Had he been more cautious and washed himself before returning home, he might have escaped detection. Had he fabricated a better story, he might, again, have escaped detection. But he did neither, and suffered the consequences. From the moment his infidelity was uncovered, his partner decided that when the right moment came she would leave him, rather than wait for him to leave her.

Had the man's infidelity not been discovered, he would probably have done better reproductively: at the very least, he would have had the chance of a third child with his partner. As it was, he did not. Nor did he manage to compensate for this loss by siring a child via infidelity or via a second longer-term partner. He nearly succeeded in fathering a child by another man's partner, but failed. He nearly attracted another fertile partner through his infidelity, but failed when it came right down to it. Any further chance he had of producing children faded when he contracted gonorrhoea (at least half of all human infertility is due to sexually transmitted disease). Even if he had remained fertile, his age and reduced circumstances made him less attractive to fertile women. Instead, his major reproductive strategy became his partnership with an older, probably post-menopausal, woman – in the sense that, with her help, his children will probably have left home fully able to realise their reproductive potential. Eventually, they would give him grandchildren. Had his infidelity not been discovered, he might have had more grandchildren, but at least he had some.

Had the woman not discovered her partner's infidelity, she would probably have done worse reproductively. Of course, she might have had another child with him. But all the time she was with him, she would unknowingly have been vulnerable to sudden desertion and decreased support. As it happened, she would also eventually have contracted gonorrhoea and risked future sterility. Instead, her discovery helped her to take advantage of his support for five more years, then pick her moment to leave him for another man and a second family. She will have ended up with the same number of grandchildren as him via their mutual children, plus more via the children in her second family.

Reproductively, infidelity has the potential to benefit both men and women, as we shall see later in this book. However, as these last four scenes have shown, it can also be costly even if it is successfully hidden from the partner. On top of this, however much a person may suffer from their own infidelity, they are likely to suffer even more if their partner is unfaithful – especially if they fail to detect it.

Sometimes the benefits of infidelity will outweigh the costs and sometimes they will not. As we noted earlier, the reproductive successes of each generation will be those people who, consciously or subconsciously, judge correctly when it is and when it is not advantageous to be unfaithful. Then, when infidelity is judged to be advantageous, the most successful will be those who hide their own infidelity while preventing or at least detecting their partner's. The reproductive failures will be those who misjudge their situation, fail to hide their own infidelity or fail to prevent or detect their partner's.

Because of these conflicts of interest, there is very little variation in the strategy shown by men, women and virtually all other species of animals which form long-term relationships – to attempt secret infidelity on the one hand while preventing a partner from doing the same on the other.

5

Secret Anticipation

SCENE 12

A Double Life

Another day at work was over, and the man on the fifth floor was just leaving for home. As he stepped into the lift, his eye was caught by a new piece of graffiti. He smiled. In colourful slang it implied that the boss of their section was not to be trusted – because he spent too much time masturbating.

The lift stopped on the third floor and a young girl stepped in. He had seen her only once before, but she had immediately won a starring role in his fantasies. She was new to the section, an extra secretary for the man who was the target of the graffiti. He watched her face as the door closed. She saw the words immediately. Her eyes flicked sideways and caught his. They smiled at each other and she blushed, but nothing was said. When the lift reached the ground floor and they stepped out, she turned right and he turned left, heading for the car park and home.

Thirty minutes later, he drove into his garage. Once in the house he kissed his partner, admired his children's latest drawings, and went upstairs to undress for a shower. Even as he was locking the bathroom door, he was handling his genitals, encouraging an erection. By the time he had shut the cubicle door and switched on the shower, his penis was ready for action. As he pumped himself, he searched through his mind for a sexual image. For about the tenth time in recent months, he tried mentally undressing his partner's sister, but the image was losing its power. He tried undressing her again, this time from the back, but it still failed. Then he remembered the young girl in the lift. OK, so the lift breaks down between floors. They both undress, he kneels down, she stands astride him, begins to crouch in order to sit on his erect penis, and . . . But there was no need for an 'and'. With that image, he ejaculated, flushed the semen down the plughole, then continued with his shower as if nothing had happened. The whole exercise had taken about two minutes.

As he showered, he mused over these routine urges to masturbate. It was as if he had a double life – sex with his partner on Saturday and Sunday, sex with himself on Tuesday and Thursday. Sometimes an occasion would be missed, sometimes an occasion would be gained, but on the whole these were his two routines.

Of course, his sex life with himself was not entirely routine. There had been the occasional highlight, particularly when he was younger. A goodnight kiss with a new girl, or a heavy but unconsummated petting session at a party, had both at one time or another sent him into a toilet to seek relief in ejaculation. Even now, an erotic scene at the movies or on TV would sometimes have the same effect. Such highlights apart, though, his sex life with himself was as much a routine as his sex life with his partner. Usually, just like tonight, it was barely even sexual, having more in common with urination and defecation than with intercourse and sexual excitement. There was little urgency, little excitement, and little feeling of satisfaction – just a feeling of relief.

~

Most men should be able to identify with this scene. For two-thirds, their very first ejaculation will have been triggered by self-stimulation. Over 98 per cent will masturbate at some time in their lives. Moreover, virtually all will have done so by the age of twenty. However, despite such widespread practice, few men would view their habit of shedding sperm as an important weapon in their pursuit of reproductive success. Yet this is just what it is. Scene 12 is the first of three in this chapter, each of which explores one or more aspects of the shedding of sperm (through masturbation or 'wet dreams').

How often a man masturbates depends on how old he is and how often he ejaculates for other reasons. On average, how often he ejaculates altogether (intercourse, masturbation, and nocturnally) will closely reflect his rate of sperm production. This varies from man to man, depending on the size of his testes. It also varies with age. From almost immediately after puberty to the age of about thirty, an average man produces around three hundred million sperm a day and ejaculates between three and four times a week. By the age of fifty these rates have dropped to about 175 million a day and to twice a week and, by age seventy-five to about twenty million a day and to less than once a month. If a man under thirty has intercourse three times or more a week, he will rarely masturbate. If he has sex only once a week, he will probably masturbate about twice.

Men are not the only mammals to shed sperm through masturbation. Dogs, of course, are notorious for such behaviour. One of their habits is to affectionately clasp a person around the knee with their front limbs and to rub their genitals up and down the dismayed person's leg. Albeit using other methods of self-stimulation, a wide range of other species, such as rats, mice, squirrels, porcupines, pigs, deer, whales, elephants and monkeys, are also known to shed their sperm. Just like men, the males of these species will often masturbate between routine inseminations of females.

Masturbation may not seem a very sophisticated activity, but it is. It is the means by which an active, or hopeful, man tailors his next inseminate to its likely circumstances. By anticipating what those circumstances might be, he can use masturbation to adjust the age and number of sperm he will then introduce into the potential female. Not only that, but he can also adjust what proportion of those sperm will be blockers, killers and egg-getters.

A man's body can distinguish between masturbation and insemination. The ejaculates produced are not identical. Many factors influence the number of sperm introduced during inter-course – such as how many are needed to top up his partner and to succeed in warfare, given the perceived level of risk. But, apart from a man's age, only *one* major factor – time since last ejaculation – affects the number shed during masturbation. When a man masturbates, he voids about five million sperm for every hour since he last ejaculated. This seems to be the rate at which sperm exceed their sell-by date as blockers, killers and egg-getters.

Sperm change their roles as they age. Most are killers when they are young and blockers when they are old. Killers need to be full of energy and movement and to have a cap full of lethal chemicals (Scene 7). Blockers, however, can be old. Or, to put it another way, the only use for geriatric sperm is as blockers. To block a cervical channel, a sperm needs to have just enough energy to swim out of the seminal pool and travel a little way. into the cervical mucus. Then it can simply coil its tail and sit still. It can even die if it wants. The channel will still be blocked.

Egg-getters also change with age. Important chemical changes have to take place on an egg-getter's surface before it can pene-trate an egg, changes that do not occur until the sperm gets close to the fertilisation zone in an oviduct. As a result, an egg-getter's life is divided into two phases of activity. It has its first burst of energy *before* it is fertile – this is when it surf-boards and swims to the rest area in an oviduct. Then, during its brief phase of fertility, it has a second burst of energy – when it reaches, then swims through, the zone of fertilisation. After that, it dies.

The two columns of sperm queuing to be ejaculated (Scene 4) are not just a homogenous, characterless mixture. Inevitably, those at the front of the queue, near the man's urethra, are the oldest. The ones at the back, way down in his testes, are the youngest. There may be some mixing of ages as the sperm are loaded into the urethra. Even so, when he places his seminal pool at the top of a woman's vagina, the oldest sperm tend to be introduced first and go to the bottom of the pool. Younger sperm arrive in later spurts and go to the top.

The younger sperm, being more active, will be the first to enter the cervical mucus. Older sperm leave the pool more slowly. Very old sperm cannot leave the pool at all and are doomed to be ejected by the woman in her flowback. From their position higher up in the seminal pool, the younger sperm may have to swim down through groups of older ones before they can swim up and enter the elephant's trunk of the cervix. Much depends on how deeply the cervix dips into the pool. Too many old sperm in the pool hinder the youngsters as they try to escape into the cervix.

Masturbation between intercourses usually means that a man inseminates his partner with fewer sperm than if he had not masturbated. However, the sperm he introduces are younger, more dynamic, and less hindered by supernumerary geriatric sperm. As a result, just as many, if not more, manage to escape the seminal pool and stay in the woman. Moreover, the sperm which escape are younger and more active – altogether a more efficient army.

All the same, a man needs to inject some old sperm as blockers – as many as are needed to top up the population of blockers in his partner's cervical mucus. For three days after a couple's last routine intercourse, the ageing of sperm in a man's tubes tracks the loss of blockers in his partner's cervix with exquisite precision. Given any gap between successive intercourses of from thirty minutes to three days, he can introduce into his partner the perfect ejaculate to top up her population of blockers. Once the gap exceeds three to four days, however, his sperm queue develops an excess of blockers. This happens because there are only so many channels in the cervical mucus, and once he has enough old sperm to block these, any further blockers are redundant. In fact, they are worse than redundant, for they probably hinder the escape of his younger sperm from the seminal pool into the cervix. Beyond a certain point, therefore, the man's best policy is to shed the oldest part of his column of sperm before inseminating a woman, rather than leaving them for her to eject in her flowback. This is one of the functions of masturbation.

If the gap between inseminations exceeds four days, the ideal inseminate is produced by masturbating *two days* before the next intercourse. This gives a man two sperm columns, each with about twenty million aged blockers at the front of the queue and from a hundred million to five hundred million younger blockers and killers further back. Scattered among these younger sperm are egg-getters, in a ratio of about one egg-getter to ninety-nine kamikaze types. About 10 per cent of the egg-getters are already at their peak of fertility and will go immediately to the oviduct. The remainder will reach their peak at different times over the five days after insemination.

This profile allows a man maximum flexibility in adjusting his sperm army according to the risk of warfare. If he has guarded his partner closely since they last had sex, he needs to introduce into her a full quota of blockers but only a few killers and egg-getters. So he injects all of the blockers at the front of the queue, plus a small portion of the queue behind containing the killers and egg-getters. If he has spent less time with her, so that the risk of sperm warfare is greater, he needs more killers and egg-getters. So he ships in more of each column. The number of blockers stays the same, but he introduces more of the younger sperm, the killers and egg-getters.

In order to have a two-day-old column of sperm waiting to be introduced, a man needs to anticipate his next intercourse with his partner. Subconsciously, the brain must play a major role in this anticipation. The urges that men get to masturbate are timed by their brains and bodies to achieve this gap between masturbation and intercourse. The man in Scene 12 most often masturbates on Tuesdays and Thursdays in routine anticipation of sex on Saturdays. Occasionally, he may masturbate only on Wednesdays. The result is that he rarely inseminates his partner with an ejaculate older than three days or younger than two. This should mean that, at each intercourse, the army he injects is ideal for the circumstances. More often than not, of course, an army is unnecessary: no war takes place, and all he really needs is to introduce a few million egg-getters of different ages. But, just in case his partner has been or might be unfaithful, he injects a measured number of blockers and killers.

The major reproductive problem a man encounters with routine masturbation is that there is an unpredictable element to routine sex. As we have seen, this unpredictability is generated largely by the conflict between men and women over the function of routine sex, and the problem occurs when anticipated inseminations fail to materialise (though spontaneous and unexpected seduction by his partner can also be problematic).

For example, thanks to his masturbation routine, the man in Scene 12 arrives for intercourse each Saturday night with an ideal ejaculate waiting in his tubes – a balanced mixture of blockers, killers and egg-getters. Suppose, however, that one Saturday night the couple argue, pass out through drink, or experience any one of the myriad circumstances that can prevent them having intercourse. On Sunday, the man now has in his tubes an ejaculate that is rapidly going past its best. His body then has the choice of either waiting for the next intercourse or masturbating. If he waits for the next intercourse, he will introduce in his partner an ejaculate that is on the old side of its prime, with too many geriatric blockers and many ageing killers and egg-getters. On the other hand, if he masturbates, but then unexpectedly has the chance of intercourse with his partner only an hour or so later, he will introduce an ejaculate short on blockers. Neither alternative is ideal and could put him at a disadvantage if either of those ejaculates really does have to fight a sperm war.

This problem for the man is an opportunity for the woman. One way she can manoeuvre her partner into injecting a sub-optimal army is to change their sex routine unexpectedly. If she is then unfaithful in the next few days, her lover's army will have the advantage (Scenes 6 and 26).

SCENE 13

Multiply, But Don't Divide

The evening flight had taken three hours. Arriving just after midnight, the four business colleagues, two men and two women, had taken a taxi from the airport and gone straight to their hotel. Their rooms were all on the same corridor. As they unlocked their respective doors, the man said goodnight to his three associates, then went into his room. They all had an early start in the morning.

Once unpacked, he undressed and showered. The room was very hot. Too restless to sleep, he lit a cigarette, then lay naked on his bed, staring at the ceiling. Handling his cigarette dextrously with one hand and his genitals abstractedly with the other, a kaleido-scope of thoughts began to pass through his head. Momentary images of his children flitted by – followed closely by images of his partner. Remembering their last intercourse, two nights ago, he mused over whether she had really climaxed. Or had she been faking? Perhaps her performance had been a gift to him, a memory to take on his trip. If so, it was a pity their mutual sexual act had been superseded. The following night, the one before his trip, he had slipped into the bathroom and masturbated while she was on the phone to her mother.

His thoughts turned to his companions and his strategy for the next two days. He could hardly have been luckier. The two most desirable women in the organisation were both on the trip, and he had two nights in which to try to bed them. Not for the first time since the trip had been announced, he toyed with the idea of being in bed with them both at the same time. He imagined them naked, writhing over his body, each begging him to enter her first.

As his fantasy unfolded, his penis grew in his hand and his brain whispered that maybe he would like to ejaculate. He stayed on the bed for a while, enjoying his fantasies and the growing sensation in his genitals. Then, when the time came, he went over to the wash-basin and finished himself off. Later, as he drifted off to sleep, he wondered how well the next two evenings would match his fantasy.

The following evening, back in his hotel room, he phoned home. After talking for ten minutes, he sent his love to the children, blew his partner a kiss, and rang off. Reassured that everything was as it should be with his family, he went to his fifth-floor window and looked out over the city skyline. Perhaps his wasn't a bad job, after all. Business was over for the day. Now all he had to do was look forward to the evening with his three colleagues. There would be food, drink and, with any luck, sex.

The evening began well but ended badly. In the taxi to the restaurant, he sat next to the woman he most wanted to see in his bed and interpreted her body language favourably. However, as soon as they hit the city bars, it was obvious she preferred the other man, his boss. As the evening progressed, he could do nothing but watch as they became more and more physical. He made a few vain attempts to get between them but she virtually ignored him. Trying to make the best of the situation, he turned his attention to his second choice. But after a few drinks she began complaining of feeling tired and unwell. He suggested accompanying her back to the hotel, hoping the evening might not yet be over – but it was. At the hotel, she made her excuses and went straight to her room. He stayed in the hotel bar for a while, then went to his own room.

All he had left for the evening were his fantasies. For the third night running, he masturbated before going to bed. It was like being a teenager again, masturbating nearly every day for a week or so at a time. He slept only fitfully. At two in the morning he heard his boss and the woman come back and, as far as he could tell, both went into his boss's room. At five he thought he heard her leaving.

The following evening was quite different. The boss ignored his conquest of the night before and made immediate overtures to the other woman. Without hesitation, she responded warmly. He tried to capitalise on the turn of events, but for a while the jilted woman was single-mindedly sullen. Then suddenly, almost as if a switch had been flicked, she began to respond to his advances. Eventually, he realised that her responsiveness had more to do with making his boss jealous than with genuine interest, but he made the most of it. She became very drunk and by the time the four of them arrived back at the hotel, he thought she was either going to pass out or throw up. He was undecided whether he should try to have sex with her, but she made the decision for him.

Waiting until the boss and the other woman could see what was happening, she grabbed his arm and virtually dragged him into her room. Once in, she fell on to the bed and began a verbal assault on their two companions. Her tirade scarcely faltered, even when he undressed and then entered her. As he had masturbated three times in the last three days, and as he was fairly drunk, intercourse was long. He needed many minutes of thrusting before he began to feel like ejaculating. His concentration wasn't helped by the way the woman scarcely seemed to know he was inside her. Throughout his sexual performance, she never once stopped talking.

After he had ejaculated, she gradually ran out of verbal steam and fell into a drunken sleep. Two hours later, he tried to have sex with her sleeping body. But she woke and, with irritation in her voice and manner, told him to stop and leave. She wanted to sleep – but first, she had to be sick. In the end he gave up, got out of bed, and went back to his own room. He was just in time to see the second woman closing the boss's door and crossing the corridor, naked buttocks peeping out from the clothes she was clutching round her front.

The four missed each other at breakfast and did not meet until it was time to take a taxi to the airport. There was an uneasy politeness, and nothing was said of the night's events. At the airport, the man bought toys for his children and perfume for his partner. The woman whose bed he had shared the night before spent most of the flight in the toilet, being sick.

That evening, back at home, he told his partner he thought his boss had had sex with both of the women. Responding to her questioning, he denied that he had found either of them attractive. Later that evening, unusually for a Thursday, he and his partner had sex, and over the next few days life at home and at work returned to normal.

Two months later, the woman who had slept with both men on the business trip announced her pregnancy. There was some story about being sick and her pill failing, but the man never quizzed her too closely. Six months after the business trip, he met the woman and her partner at a party. As he listened to the certainty with which the couple spoke about *their* forthcoming baby, even he began to wonder if he had really ever had sex with her.

∽

The role of masturbation in a man's pursuit of reproductive success is not limited to its role in routine sex (Scene 12). In fact, in many ways, it becomes a much more potent weapon in his preparation for infidelity and sperm warfare.

As we discussed in Scene 12, most of a man's sex life with

himself is just as routine as his sex life with his partner. Not surprisingly, as routine sex is the main focus of his sexual activity, it is this to which his body is most finely adjusted. But infidelity introduces a non-routine element into his sex life with his partner – and, therefore, into his sex life with himself as well.

When a man has an opportunity to inseminate a woman other than his partner, he has a problem. The ejaculate waiting in his sperm tubes is tailored to top up his partner, and so is relatively rich in blockers. What he needs for infidelity, however – if it is to be reproductively successful – is an ejaculate rich in killers and egg-getters. The woman he is about to inseminate has an above average chance of containing sperm from another man, especially if she has a regular partner. She will already have blocking sperm in her cervix. These may or may not need topping up, depending on how recently she has had sex, but blockers are not a lover's main concern. What he needs is young, active sperm that can get through the cervix in numbers and prevail in combat. Blockers are at best a luxury; at worst a hindrance. As with the man in Scene 13, once the brain begins to anticipate infidelity, the body increases the masturbation rate, thereby producing and maintaining a young, killer-rich ejaculate in his tubes, ready and waiting for action. The ideal inseminate for infidelity is one that has been waiting to be ejaculated for no more than about twenty-four hours.

Another situation in which the ideal inseminate is twenty-four hours old or less occurs when a man has no regular sexual partner. (All men have such a phase during adolescence, and many have additional such phases later on when 'between partners'.) In this situation too, he needs to keep his sperm column young and in a state of continuous readiness – intercourse and sperm warfare could be just around the corner. As he searches for a sexual opportunity, he needs a waiting ejaculate that is heavily dominated by young, energetic killers and egg-getters primed to head straight for the oviducts. Frequent masturbation maintains such an ejaculate – hence the appropriate urges experienced by single men.

Almost as interesting as the function of masturbation is its secrecy and lack of social acceptability. Among humans, it is most often conducted in private. It is often also the target of prejudice. The graffiti artist in the lift in Scene 12 has his counterparts all over the world. Everywhere, people vilify men for masturbating, often using terms of abuse specific to the behaviour. So strong is this social disapproval that many cultures and religions actually outlaw the activity. Even in cultures that stop short of outlawing it, stories and folklore may exist (such as 'masturbation leads to blindness') that are designed to scare men into refraining. The irony, of course, is that the graffiti artist and all his counterparts world-wide will almost certainly practise masturbation themselves. So why *is* masturbation so often surrounded by secrecy, prejudice and hypocrisy?

If a man is to hide infidelity from his partner, it is important that there is no overt change in his behaviour: any change could alert her to his new situation. He cannot, for example, alter their usual sex routine. Nor – if she is aware of it at all – must his partner notice a change in his masturbation routine. As we have seen, a sudden increase in his masturbation rate could mean he is anticipating infidelity. A sudden decrease could mean he is already being unfaithful, ejaculating so often through intercourse that he would gain no benefit from masturbation.

Assuming, again, that he had always been secretive about his masturbating, he has two options. The first is to maintain his usual routine. But if he does this, he injects inefficient sperm armies into his lover. The alternative is to continue to be generally secretive over his masturbation so that no change in routine can be noticed. It is, of course, no threat to a man's strategy if his partner detects an occasional episode. In fact, deliberately allowing detection can sometimes be a strategy in itself – especially if it reassures or misleads. A man need worry only if his partner detects so many occasions that she is able to work out his whole masturbation routine.

Obviously, secrecy is most important once males have partners, or are competing for partners. Then, masturbation patterns need to be hidden not only from the partner but also from other males. Imagine a situation in which a man allowed a male friend to know his routine. Then his body urged him to increase his masturbation rate in anticipation of infidelity – with the friend's partner! Just in case such a situation might arise, it is far better for a man to hide his masturbation routine from everybody in the first place. Only among groups of adolescent males without partners is such secrecy perhaps less important. All adolescents know that they are each on the lookout for intercourse. Frequent masturbation, even if obvious to other partnerless males, gives little away.

The secrecy which surrounds masturbation is therefore understandable, given its function in the context of reproductive success. So, too, is prejudice and hypocrisy. Since the aim of masturbation is to give the male an edge over others in sperm warfare, he gains most if he masturbates but can dissuade those around him from doing so. That way, *he* gains competitive benefits that his rivals do not. The world-wide tendency to criticise, even victimise, other people for masturbating while continuing to masturbate oneself is thus as strategic as masturbation itself.

There is one further element in Scene 13 that should be discussed now – alcohol. This is, and has been (probably for many thousands of years), an important factor in many people's sex lives. It also has a very well documented effect on sexual behaviour. The more men and women drink, the more they both seek intercourse – or, at least, the less they resist it. Perversely, however, beyond a certain level of intoxication, both men and women find it more difficult to become aroused. Men find erection more difficult – the 'brewer's droop' syndrome – and women find their vaginas, and often their clitorises, less sensitive. This secondary effect eventually makes intercourse impossible for a man, but not for a woman. Thus, the more men drink, the more they want sex but the less able they are to have it. In contrast, the more women drink, the more they want and can have sex. It is just that, beyond a certain point, they may not enjoy it much.

Alcohol does not negate any of the strategies we discuss in this book. All of the patterns are still there: for example, women still show a menstrual cycle of varying sexual interest. Even under the influence of alcohol, they are still more likely to be unfaithful during their fertile phase and still more likely to have sex with their partner during their infertile phase. The only difference is that, under the influence of alcohol, intercourse at all times is more likely to occur.

In Scene 13 we have no way of knowing who in fact fathered the woman's child. At the time she ovulated, she probably contained sperm from three different men – her partner, her boss and the main character. Her partner had a two in three chance of being tricked into 'fathering' another man's child. For the purposes of this discussion, let us assume that this did happen and that the father was the scene's main character. This man will thus have achieved the primary reproductive advantage of infidelity (for a male): he will have multiplied, increasing his number of children. Yet he will not have divided his relationship – there will be no price to pay from being deserted by his partner. Moreover, having successfully tricked another man into believing that he is the father of a child that is in fact not his own, our main character will not be called on to help support the child. In short, he will have reaped the reproductive benefits of infidelity, but suffered none of the costs.

If the man's infidelity did give him a genetic child extra to those with his partner, he probably owed his gain to his secret preparation. His success may well have been due to those three masturbations and the resulting potency in sperm warfare of his young and deadly army.

SCENE 14

A Wet Dream

The man stirred in his sleep as his partner turned over in bed. Briefly her hand brushed his erect penis. His nightmare was interrupted, but only for a second. The shower cubicle had opened into the street and he had gone out by mistake. It was early morning, still dark, and there were few people to see him as he wandered naked around the streets, looking for the way back to his shower. But two or three had walked past him, pointing. He could hear them laughing, but couldn't see their faces. Then a gang of lads with knives saw him from across the street. They chased him, shouting that they were going to castrate him. He ran and ran, breathing more and more heavily, wishing his penis would lose its erection so that he could run faster.

Suddenly he was at work, being told by his boss that if he wished he could leave his clothes off, but his erection had to go. An anonymous, faceless woman came into the room with a hat in her hand.

'Cover it with this,' she said and reached forward to do it for him. But she dropped the hat and grabbed his penis instead, squeezing so hard it hurt.

'We can't waste that, can we?' she said, and in an instant was naked.

The room filled with fog. All he could see was the woman's body from her waist to her knees. Her black pubic hair filled his tunnel vision and as she lay down, pulling him on to her, she became just a vagina. But he couldn't get into her. Wherever he pushed his penis, there was a barrier. Where was her entrance . . .

where was it? Panic set in. He was going to ejaculate before he was in. And he did.

The woman disappeared instantly. So did his panic. But he didn't fully wake until he felt the semen trickling over his belly, down his side, and on to the bottom sheet. He groaned inwardly and wiped the ejaculate off his body with the top sheet.

That was his first wet dream for a year. He turned over, away from the rapidly cooling wet patch. Briefly, he was an adolescent again, worrying how he could hide the patch from his mother in the morning. At times during his early teenage years, he had even worried about going to bed at all in case he ejaculated. It was bad enough at home, but if ever he stayed at a friend's house he could scarcely sleep all night for fear of leaving his mark. Eventually, he had discovered that if he masturbated before going to sleep when away from home, wet dreams rarely happened.

Since then, these nocturnal surprises had virtually disappeared from his life. Now, in his early thirties, they were a rare event. The last time, a year ago, was when he had flu and a slight fever. Then his partner had reacted badly on discovering the mark on the sheet, claiming it was a slight on her sexual attractiveness – and how could he be feeling like sex if he was ill, anyway? He wondered if she would react the same way this time. But, unlike during adolescence, the thought of being discovered did not keep him awake. Within a minute, he had drifted back into sleep.

～

For one in five men, their very first ejaculation occurs spontaneously. Most often, this first occurrence is a wet dream, but for a few unfortunate boys it occurs in public. This is usually due to subliminal stimulation (watching a film, climbing a tree), or extreme stress (reading in front of a class at school). After the first few occasions, however, almost all spontaneous ejaculations occur in private, at night, while asleep – as wet dreams. At some time in their lives, over 80 per cent of men experience these nocturnal events. So, how do they help a man in his pursuit of reproductive success?

Wet dreams can occur at any time of life, but are most common in the teens and early twenties, and are most likely after periods of abstinence from intercourse and masturbation. But only during the teens do they occur often enough to be a direct substitute for other outlets. The adolescent's body aims to ejaculate at least three times a week, and if he omits to do so via masturbation (or intercourse, if he is lucky), wet dreams act as insurance.

During adolescence, wet dreams probably tailor the ejaculate just as precisely as masturbation. Later, they occur much too rarely to be part of the three-a-week system. From his mid-twenties onwards, even if a man fails to ejaculate via intercourse or masturbation for days or even weeks, wet dreams may not occur. He will still get virtually nightly erections during REM sleep (the deepest kind) and dreams, but his body seems to inhibit ejaculation.

After adolescence, nocturnal ejaculations are probably most likely to occur when the sperm waiting to exit have been damaged in some way. A high temperature, for example, during flu or other infections, can kill or damage these waiting sperm. The body still generates an urge to masturbate under such circumstances, but if the man gets no opportunity for secrecy during the day, the body forces a wet dream to rid his sperm columns of these moribund tenants at night.

Humans are by no means the only species to ejaculate spontaneously. Rats, cats and probably most mammals will from time to time ejaculate in their sleep, especially around puberty. In men, it is sleeping with a partner that most often seems to mark the change from the function of wet dreams as a back-up in his three-a-week system, to their being rare events.

As we have already discussed (Scene 13), the timing and frequency with which a man sheds sperm can give a clear indication to an observer both of his current sexual activity and of his expectations. Secrecy is important to the success of male strategy, and it is probably this need for secrecy that has caused men's bodies to avoid wet dreams as sexual outlets once they have a partner. They are not as easy to hide from a partner as masturbation, because the evidence remains on the sheet. Since a man has much more control over when and where he ejaculates during masturbation, wet dreams are a suitable back-up for masturbation only in young men without a partner.

6

Successful Failure

SCENE 15

Home for the Day

The soldier stopped pedalling and allowed his bicycle to free-wheel down the steep hill. He welcomed the brief moment of relaxation. It was over two hours since he had cycled out of his barracks in the heat of the June evening. Now, as the sun dipped towards the horizon behind him, he still needed another half-hour to get home. His bike had no lights, but he might just arrive before it would matter. As he gathered speed and the fields and hedgerows raced past him, he could scarcely believe his luck, both bad and good. The bad luck was that, for some wartime reason, the whole barracks was being moved much further away from his home. The good luck was that he had been given an unexpected twenty-four-hour leave. He had his bike, he had the strength and energy, and he certainly had the motivation. A three-hour ride? Nothing to it.

His partner didn't know he was coming home. For a moment she was stunned when, just as it was getting dark, the back door opened. Her two daughters, aged eleven and seven, were upstairs, asleep. Otherwise, she was alone in the house. When she identified the intruder, she was both pleased and unnerved. Pleased, because she missed his company during the long weeks between visits home. Pleased, because she had been feeling sexually frustrated for days. Unnerved, because if he had arrived an hour earlier he would have found another man in the house.

So far, her genuine claims of fidelity during his long absences had always been accepted – eventually. Her position in such pleadings was particularly vulnerable because her elder daughter was not his, and he knew it. She was the product of a springtime indiscretion. As a twenty-two-year-old farm girl, she had allowed herself to be coerced into sex with the man for whom she worked. Never would the woman forget that day, twelve years ago, that nearly ruined her life. Even now, the man's eyes still stared at her in her dreams and from her daughter's face.

On finding she was pregnant, she had lost her job, her mother's love, and her freedom. She was banished, first to live with her grandmother for a few months during the most obvious phase of her pregnancy, and then to a confinement in the house of a tyrannical, moralistic midwife. There was no doubt that she owed this soldier who had just walked through the door a lifetime's gratitude for offering to take on and support her and her baby daughter. And he had done this despite the opposition of his own parents and despite the prejudice of the villagers in whose community they lived and worked. Four years after they had met, she had given him his own daughter, the consequence of New Year celebrations – but he desperately wanted a son. Maybe if war hadn't broken out, they would have had another child. But it did, and they hadn't.

He enjoyed the camaraderie of army life, but constantly worried about what his partner was doing back at home. Knowing this, she shuddered at what might have happened had he arrived home and found another man in the house. In fact, her relationship with this other man had so far been totally platonic. But she knew his intentions were far from innocent and tonight, for the first time, she had found herself being tempted. Who knows what would have happened if the girls had gone to bed at their proper time, instead of staying up as unwitting chaperons?

After a welcoming hug and kiss, she told her partner to wash away his cyclist's sweat while she prepared what food she could from a temporarily thin larder. He waited until they were in bed and becoming aroused before breaking the news to her that he'd had no chance to buy or borrow any condoms. She protested that the last thing she wanted at the moment was a baby. He said it would be all right because he would withdraw before he came. Maybe it didn't matter anyway, she said, because her period was due any day.

He didn't withdraw, either then, first thing in the morning, or again in the early afternoon. An hour later he was gone, cycling the three-hour journey back to his barracks and imminent relocation. Even as he cycled, one of his sperm was entering his partner's egg, and by the time he saw her again she was experiencing pregnancy sickness. A few months later, during the hottest week in March for twenty years, his son was born.

Fifty years later, the woman would have seven grandchildren, the soldier six – one from his daughter, and five from the son conceived when he was home for that day in June.

~

Each of the three scenes in this chapter deals with a situation in which a different woman appears in some way to fail. Yet in failing, each actually benefits – as if her failure were strategic. In Scene 15 we encounter a woman who is apparently convinced that the last thing she wants is to conceive. However, not only does she fail to avoid conception, but her body goes to some lengths to make sure that she does. Fifty years later it is clear that her body had engineered a very successful failure – a real bonus in her pursuit of reproductive success.

The conscious reason that this woman allowed unprotected sex, even though she didn't want to conceive, was because she thought conception was unlikely. Her spur-of-the-moment reasoning was that it was so long since her last period that her next must be due any day. But as we noted in Scene 2, and as she soon discovered, the human menstrual cycle is nowhere near as predictable as many people think.

Just about the only reliable element in the cycle is that once a woman has ovulated she will menstruate fourteen days later (but see below). Everything else is highly variable, as her body contrives to make itself unpredictable to men. Particularly crucial to her strategy is the tremendous variation in the number of days from the end of menstruation to ovulation. This variation (which we shall soon discuss further), however, is only one element in her unpredictability. Another less well-known feature is that even apparently normal cycles are by no means always fertile: a woman experiencing such cycles may not in fact be ovulating at all. Scattered among her fertile cycles, every healthy, fertile woman routinely also has infertile ones.

There are at least three different types of infertile menstrual cycle – and most women experience them all. In the first type, a woman neither menstruates nor produces an egg. In the second, she menstruates normally, but still does not produce an egg. In the third, she both menstruates and ovulates, but curtails (from fourteen days to about ten) the normally consistent interval from ovulation to menstruation. This curtailment prevents a fertilised egg from implanting.

In the short term, phases of infertility are an important method by which a woman's body subconsciously seeks to do two things: on the one hand, to confuse men; and on the other, to organise the number and spacing of children over a lifetime. Only if such infertile phases persist for long periods, such as several years, might they indicate a problem. Even then, they may still be part of a woman's natural subconscious ability to plan her family.

The proportion of cycles that *are* fertile varies in a fairly predictable way throughout a woman's life. Obviously, she does not ovulate during childhood, though she may ovulate before having her first period. And even once she has begun to have periods, few cycles are fertile. By the age of twenty, a normal, healthy woman produces an egg in fewer than half of her cycles. Even at her peak, when around thirty years old, still only about 80 per cent of cycles are fertile. After thirty, the proportion begins to decline, slowly at first, then rapidly after she reaches forty. By fifty, women more or less cease to ovulate, though there are unconfirmed reports of women giving birth up to the age of seventy.

What seems to happen in each cycle is that in the days after the beginning of a period, a woman's body goes through a series of hormonal changes. These changes prepare her body to produce an egg but, about one or two days before ovulation can occur, her body goes on hold. Whether she eventually produces an egg or not depends on the events of the following few days, or even weeks. This holding period is an opportunity to collect sperm, perhaps just from her partner, perhaps not – perhaps just from one man, perhaps from two or more. In part, whether or not she ovulates will depend on how her body feels about the man or men from whom she has collected sperm. Most of all, however, it will depend on how it feels about trying to produce a baby in the current circumstances.

The woman in Scene 15 thought her next period must soon be due because it was so long since her last one. She was wrong. Her body was on hold. In the absence of her partner for the foreseeable future, it had momentarily considered collecting sperm from another man. But when it collected sperm from her partner, it responded by producing an egg. Consciously, the woman thought it was a bad time to conceive, but her body knew better. Given that it was eight years since she had last conceived and that her ability to have and raise a third child would soon begin to decline, now was precisely the time to conceive – especially as it was June.

Many mammals, such as squirrels, sheep and bears, breed only at a particular time of year. This way, their bodies can time giving birth and raising young to coincide with the best weather and the greatest availability of food. In contrast, most of the larger monkeys and apes, living in more uniform tropical environments, will breed at any time of year. Nevertheless, conceptions and births are not evenly spread throughout the seasons: more still occur in some months than in others. Humans are just the same.

At the latitude of Britain and Canada, more babies are born in early spring (February and March) than at any other time of year. There is a second, smaller peak in early autumn (September). At the latitude of Central America, more babies are born at the coolest time of year (December/January). In the southern hemisphere, the birth seasons are the same as in the northern. But the months, of course, are half a year different.

Just as these seasonal birth peaks reflect peaks of conception nine months earlier, so do these peaks of conception reflect peaks of ovulation. Suppose Scene 15 was set in Britain. There, women are more likely to produce an egg in May/June and December than in any other month of the year. If the scene had been enacted in October rather than June, the outcome might have been very different. In October, even if the woman's body had been on hold, she might not have responded to intercourse by ovulating. But it was June. Moreover, she was thirty-four years old, it was eight years since she last conceived, and her body rarely had the chance to collect sperm. Despite what her brain might have thought, her body decided on that summer day that it was actually a very good moment to conceive – and it was right.

Fifty years later, her son had not only survived, he had given her five grandchildren. The average number for women living at her time and in her society is likely to have been four. Without her son, she would have had only two grandchildren, half the average. Because she had had her son, she achieved seven grandchildren in all, nearly twice the average, giving her every chance that her descendants would be present in the generations to come.

The man, of course, did not fare quite as well as the woman from their partnership. But he still fared quite well, and for him the son was even more important – without him, he would have had only one grandchild. Then there would have been a very real danger that accident or infertility would rob him of all descendants within a generation or two (Scene 1). Because he had had his son, he too achieved an above-average number of grandchildren. So from the point of view of his reproductive success it had been a good decision to take on this woman, despite the cost of raising her daughter from another man. It had also been a good decision to take on a three-hour cycle ride in the heat of that June day.

SCENE 16

The Stress of It All

As the woman fingered the piece of paper in her hand, tears began to roll down her cheeks. Ironically, only yesterday she had thought that things couldn't get any worse.

Seven years ago, when she and her partner had first started living together, the future had looked bright. But, somehow, it had all gone wrong. Maybe they shouldn't have overstretched themselves financially with their first house. Maybe they shouldn't have spent so much on holidays and entertainment in the first few years together. Maybe they should have seen the danger signs earlier and drawn in their horns when they began to slip into debt. But her partner had insisted they could manage. Promotion and solvency were always just around the corner. But now, seeing this final demand, she realised things were never going to get better. As she sobbed, it seemed to her as if life had been one long chapter of stress and unhappiness.

It had begun ten years ago. In her late teens she had become involved with a brilliant young man. But their relationship was fraught. He had been continually critical of her body and general appearance and had refused on principle to use, or allow her to use, any forms of contraception other than rhythm and withdrawal. Haunted by the fear of pregnancy and obsessed by the fear of becoming fat, she had lapsed into anorexia. After a year of despair, she finished with him, counselled herself out of her eating problems, and eventually qualified with distinction for the career she had always wanted. Two years later, she met her current partner and they began living together almost immediately.

For four years, they had always been able to find an excuse to wait 'just another month or so' before having a family. More than anything, they were waiting for his promotion – but it never came. Eventually, despite their financial problems, they decided to wait no longer. Part of her motivation had stemmed from her secret hope that a family would draw them back together again – because as their debts had mounted so their relationship had deteriorated. More and more frequently, irritability had begun to give way to open hostility.

To her dismay, she hadn't conceived. Month after month went by, and she began to worry that she was infertile. As it turned out, her fears were unfounded. But it took nearly a year of unprotected sex before, relaxing on a holiday they couldn't afford, she eventually conceived. Then, nearly three months later, her partner lost his job and within a week she miscarried.

From that tragic moment, their lives had gone from bad to worse. The house was repossessed and they were forced to move from one rented accommodation to another. With each move, as they struggled to manage on her income, they lowered their sights and their standard of living. Their current flat was cramped. Although pleasant enough in summer, it was cold, damp and mouldy in winter. Eventually, her partner had got another job, less well paid than the first but at least with prospects. By then, however, they were so far in debt they had no choice but to stay in their cheap flat, waiting for financial recovery.

During the worst of their deprivations they rarely had sex, and whole months sometimes passed without intercourse. With her partner's new employment, however, their sexual interest in each other returned. But it was now two years since her miscarriage, and still she had not conceived. Their finances were slowly improving, but their standard of living was still far worse than they had originally expected it would be by now. They still argued frequently, sometimes violently, and often had bouts of ill-health.

Suddenly, she stopped crying, screwed up the paper and threw it at the wall. Her mind was made up. She scribbled a note to her partner and walked out of the door. At the end of the street, she went into a telephone kiosk and dialled a familiar number. When the man answered, she gave him the message he had been waiting all month to hear: if his offer was still open, she was ready to leave her partner and move in with him.

Ten minutes later, she was in his car. Thirty minutes after that, they were in his house in the suburbs. Ever since their affair began, he had been trying to persuade her to move in. His house was nothing special, but compared to the pit in which she was living it was a palace. Although maintenance payments to his ex-partner were a drain, he was not in debt, his house was warm, dry and well appointed, and he had a car.

She spent most of the day in tears and most of the night having sex. The next morning, she went back to her flat to collect her belongings. Her new partner wanted to accompany her, but she wouldn't let him and took a taxi instead. Her ex-partner was in bed when she arrived. As she collected her clothes and other pos-sessions, the arguments and recriminations reverberated around the bedroom. He eventually collapsed in tears. Then, begging her not to go, he told her how much he still cared for her.

She responded in a way that later she still could not believe. As he sobbed, she suddenly remembered the athletic, ambitious and arrogant young man she had once found so attractive. In a flood of compassion, she comforted him, calmed him, then positively encouraged him to have intercourse with her. However, almost immediately afterwards, she turned on him again, announced it was still over, and left.

Over the next few weeks, she settled into her new relationship. Sexually, they were very active – until she began to feel sick every morning. Having told her new partner she was infertile, it was a surprise to both of them to discover she was pregnant. They had no idea when it had happened until a routine scan told her the baby's age. Then they calculated that she had conceived the very first week they had begun to live together.

Her ex-partner refused to leave them in peace, and from time to time throughout her pregnancy caused great stress. The baby threatened to miscarry on several occasions, but this time she hung on to it. In due course she gave birth, albeit slightly prematurely, to a small but otherwise healthy daughter. Now her ex-partner became even more demanding, claiming the child to be his. Surrounded by conflict, the woman lapsed into a deep post-natal depression and began to neglect and even to mistreat the baby. It might have died had her new partner not devoted himself to its protection and care, losing his job in the process. The woman returned to work, leaving her new partner to take the major role in raising the child.

Her ex-partner faded from their lives for a while. The woman emerged from her depression, her new partner found employment again, and they were able to afford child care. Then, just as her life was settling into a comfortable routine for the first time in years, her ex-partner reappeared. Now with a good job and rejuvenated physically, mentally and financially, he wanted to press his claim for the baby he insisted was his daughter. He paid for a paternity test. The new partner cooperated, confident that the results would end the argument once and for all. They did, but not in the way he had hoped. He was devastated. The baby girl he had cared for so much was not in fact his daughter.

Within weeks of the result, the woman left to share her old partner's new and much improved lifestyle, taking her daughter with her. From then on, the woman's fortunes went from strength to strength and within three years of their reunion she and her original partner had two more children.

~

In Scene 15, we met a woman who enhanced her reproductive success by conceiving when, consciously, the last thing she wanted was another child. In this scene we meet the converse – a woman who enhances her reproductive success by failing to conceive when consciously she really wants to.

Most people think of family planning and contraception as modern inventions. They are not. Even 'deliberate' contraception is not that new. For centuries, women in various cultures have been placing leaves or fruit (even crocodile dung!) in their vaginas in an attempt to avoid conception. Chemical contraception, as in the pill, isn't even a human invention. Female chimpanzees, for example, at appropriate times chew leaves that contain contraceptive chemicals. In fact, the female body was planning families and avoiding conception for tens of millions of years before humans had even evolved. Women simply inherited these natural traits from their mammalian ancestors.

For women and all other mammals, the natural mediator between bad conditions and the avoidance of reproduction is stress. The stress reaction usually *seems* like an enemy – a pathological condition that we cannot shake off and that prevents us from functioning efficiently and normally. But there is another interpretation: that the stress reaction is a friend – a means by which the body prevents itself from doing anything disadvantageous when times are bad. In particular, stress is a powerful contraceptive. And, to a woman, the avoidance of conception is an invaluable ally in her pursuit of reproductive success! How can this be?

The key paradox is that a woman does not necessarily achieve greatest reproductive success simply by having as many children as possible as quickly as possible. The very fact that women, like most primates, usually have only one child at a time is testimony to the dangers and difficulties of trying to raise more than one child simultaneously. Twins might seem a good way of increasing reproductive success, but there is more than twice the danger that both will die. Unless the woman's circumstances are very favourable, she is less likely to raise two healthy, fertile children by conceiving twins than she is by conceiving two children a few years apart.

Women inherited a basic problem from their primate ancestors – it is very difficult to carry more than one child at a time when walking long distances. This difficulty is particularly great when walking upright on two legs, and has plagued women throughout human evolution – it is not totally unfamiliar even to those living in modern industrial societies. Of course, the limitation was, and still is, particularly crucial in those cultures in which women are responsible for collecting and carrying large quantities of food, water, firewood, or other materials. Carrying even one child as well is difficult. Under such circumstances, the greatest reproductive success is achieved by those who avoid having another child until the previous one can not only walk but can keep up.

The time interval between successive children is not the only aspect of 'family planning' that influences a woman's reproductive success. Children are most likely to survive and grow into healthy, fertile adults if they are born into a favourable environment. Plenty of space and an adequate supply of healthy, nutritious food are paramount. Children then have the lowest risk of contracting diseases and the greatest resistance to those they do get. In modern societies, space and nutrition depend on wealth. Even now, the chances of a child from a poor family dying before reproducing are double those of a child from a rich family. In the historical and evolutionary past – when wealth will have been measured not in terms of money but in terms of crops and livestock, or even simply in terms of access to the best areas for food, water and shelter – these differences will have been even greater.

There is a general principle, considered briefly in Scene 8 in connection with the benefits and dangers of a 'one-child' family: namely, that for any given woman in any particular situation there will be a size of family which will give her the greatest number of grandchildren. If she has a smaller family than this, she will naturally have fewer grandchildren. Equally, if she tries to have a larger family, she risks an overcrowded household and spreading herself and her resources too thinly. The resulting disease and infertility again mean that she ends up with fewer grandchildren. The challenge that faces every woman is first to identify the optimum family size for her circumstances, then to ensure that she has that number of children. Another factor that influences the number of grandchildren a woman may have is how she times her conceptions. Life is a mosaic of situations. Health, wealth and circumstance vary with time, and some periods in a woman's life are better for having children than others. Those who time their conceptions to coincide with the good times will produce the most grandchildren.

It is not just the children who may suffer from being conceived when times are bad – the mother may suffer also. An untimely conception may damage her health and situation so much that she may never again be able to conceive. If the woman in Scene 16 had produced a child during any of the most stressful phases in her relationships, she could have suffered irrevocable damage (the demands of a child when a couple can scarcely maintain themselves may even be fatal). Under such strained circumstances, there is an increased risk of ill-health and subsequent infertility. Hostility also increases, making physical abuse or even murder more likely. The woman in the scene, by delaying her first child until circumstances improved, managed, eventually, to have three children for whom she had the space, time and resources. In the end, she also had the support of an able partner. Had she tried to reproduce earlier, she might not only have failed but might also have forfeited her chances of conceiving in the future.

The best method of avoiding conception is, of course, sexual abstinence, and the woman and her partner did indeed have spells without sex in their most stressful days. This happened not because they consciously wanted to avoid conception – on the contrary – but because they lost interest in each other. At times, they even felt hostile. Their bodies were manipulating their emotions to reduce the chances of conception. Rarely, however, did they give up routine sex for long. This is because abstinence is in general disadvantageous as an overall contraception strategy – for the following reasons.

As we discussed in Scene 2, the primary function of routine sex is not conception: it is for the woman to confuse the man – and it is for the man to protect himself against his partner's infidelity by maintaining a sperm army inside her. Since neither partner can afford to give up routine sex for too long, both men and women have mechanisms other than abstinence which, despite continued sexual activity, reduce the chances of conception when the situation is unfavourable.

Women in particular have a wide range of such mechanisms. One of these is widely known: the influence of lactation on ovulation. If a woman breast-feeds her baby for a few months after giving birth, she is unlikely to ovulate during that time. This is the case even if she resumes her periods. Absence of ovulation is one of the main ways in which a mother spaces the conception of successive children.

However, most of the ways in which she avoids reproducing at inappropriate times involve stress. In Scene 16, not only did the woman avoid conception during her most stressful phases, she also miscarried, then later threatened the life of her new-born baby, in response to stress.

The stress reaction manifests itself in many ways. In the scene, faced with a difficult relationship during adolescence, the woman lapsed into anorexia, as do 1 per cent of girls between sixteen and eighteen years of age. The physiological stress caused by such near-starvation is contraceptive – ovulation, and often menstruation, are inhibited. Usually the situation is temporary. Although a few anorexics (5–10 per cent) die as a result of their behaviour and a few more (15–20 per cent) continue to be anorexic throughout their lives, the majority (75 per cent) emerge from the condition to live a normal, healthy and eventually reproductive life.

Most contraceptive reactions are less extreme than anorexia. Even so, the more a woman is stressed the less likely she is to ovulate (Scene 15). She is also less likely to help sperm reach the egg or to allow a fertilised egg to implant in her womb. Finally, she is more likely to miscarry, particularly during the first three months of pregnancy.

It is estimated that, whereas most fertilised eggs survive to reach the womb, on average about 40 per cent fail to implant and, of the remainder, about 60 per cent die before the twelfth day of pregnancy. Even then, about 20 per cent are miscarried during the next three months. All of these figures are higher if a woman is stressed, lower if she is not. The death of a partner, a partner's infidelity, or the outbreak of war, for example, are all known to increase the chances of miscarriage. During the first few months, miscarriage is also more likely if there is anything wrong with the baby, either genetically or developmentally.

At first, it might seem odd that a woman has so many different ways of avoiding having a child. Certainly, if any one of these systems were efficient, there would seem no need for others. But this apparent excess of responses is not an error in female programming. Circumstances change, often quickly, and a woman's body needs to respond equally quickly. For example, circumstances may be favourable when she ovulates but no longer so by the time the egg reaches her womb. So she will ovulate, but then avoid implantation. Or circumstances may be favourable at the time of implantation but may become unfavourable a month or so later. So, having become pregnant, she miscarries.

Even if circumstances remain favourable throughout early pregnancy, they may yet deteriorate before the baby is born. The last three months of pregnancy are often associated with marked changes in a woman's psychology. First, there are the well-known spells of 'nest-building' – strong urges to prepare the environment into which the baby will be born. Also there are spells of intense reappraisal – primary targets are her partner, home and general environment. These spells often manifest themselves as phases of worry, depression and irritability. Finally, there is often a preoccupation with the future. Any major deterioration in a woman's circumstances at this time can lead to pathological depression and the later rejection, or even abuse, of the baby.

Post-natal depression as an irresistible urge to abandon, abuse or even kill a new baby is widely recognised. So widely, in fact, that many legal systems around the world accept that a woman may not be responsible for her actions in the phase immediately after giving birth.

Throughout human history infanticide has been, and still is, one of the major forms of family planning employed by women. In hunter-gatherers, people who live by hunting and foraging rather than by cultivation, about 7 per cent of children are killed by their mother. According to the World Health Organisation, infanticide was the most prevalent form of family planning in late-nineteenth-century Britain.

Such behaviour is not restricted to humans. Like all the other forms of natural family planning, we inherited infanticide from our mammalian ancestors. Anybody who has kept pets such as rabbits, gerbils, hamsters or mice will know that if the mother is at all stressed soon after giving birth, she is likely to kill, even eat, some or all of her litter. Such infanticide is not pathological: it reflects the mother's subconscious decision not to raise the litter in the current circumstances. She opts instead to delay her attempt until circumstances improve.

So far, we have concentrated on family planning from the woman's viewpoint. But men, too, have many of the same problems. They direct a large part of their reproductive effort into preparing an environment for their children. During hard times, a man should want to avoid raising children just as much as his partner.

Most of the time, a couple's interests coincide. The woman's body will plan their family to their mutual benefit. But there are times when their interests do not coincide. Then, a man needs a contraceptive mechanism of his own. Interests are most likely to conflict when the couple's circumstances are almost, but not quite, good enough to have a child. In this situation, if the woman meets someone her body deems to be genetically superior to her partner, she may be tempted into trying to have a child with another man. In such circumstances, discussed further in Scene 18, the reproductive benefit to the woman of a child with this other man would be worth the risk of overstretching her and her partner's resources. Obviously, her partner would gain no such benefit.

For him, this is a dangerous and delicate turn of events. Somehow, he has to prevent the situation just described from happening. At the same time, he has to avoid impregnating his partner himself. Without knowing it, the man in Scene 16, the woman's original partner, was in this situation in the weeks before her desertion. The only option for such a man is to use routine sex to prevent his partner from conceiving to anybody, himself included. And men have a very sophisticated mechanism by which they can do just that: they wage war on their own sperm.

There are two types of sperm in a man's ejaculate which strongly reduce the chances of conception if they are present in large numbers. It seems that these sperm might actually be programmed to destroy a man's own egg-getters. One of them has a cigar-shaped head and is called a tapering sperm. The other has a pear-shaped head and is called a pyriform sperm. Just as for women, stress is contraceptive. When men are stressed, they produce much greater numbers of these so-called *family-planning sperm*.

The more family-planning sperm and the fewer egg-getters a man introduces into his partner, the lower the chances that she will conceive. But if she is unfaithful and sperm warfare occurs, her partner's family-planning sperm can help his conventional killers overcome the other man's egg-getters. Furthermore if, despite the partner's efforts, an egg-getter does get through, there is at least a chance that the child will be his. Had he responded to stress by abstaining from routine sex, there would be no such chance.

In Scene 16 this strategy worked successfully for the woman's first partner. He managed to prevent her becoming pregnant throughout her weeks of infidelity. But when, eventually, she conceived, he was the victor in the sperm war she had promoted.

As we have now seen, the bodies of both men and women have a range of natural methods, mainly triggered by stress, by which they avoid having children when times are bad. We might expect, therefore, that societies living in generally more stressful situations would have fewer children. But they do not – in fact, exactly the opposite is true. How is this possible?

The explanation is that the factors that influence the best *number* of children to have are different from the factors that influence the best *time* to have those children. Historically and geographically, the number of times a woman gives birth in her lifetime is linked closely to the chances of her children surviving to adulthood. When survival prospects are poor, a woman gives birth to many children. If she does not, few or none may survive to adulthood. She avoids having children during only the very worst times, and conceives whenever circumstances show signs of improving. In contrast, as we saw earlier, when survival prospects are good a woman gives birth to fewer children and lavishes much more of her time and resources on each. Her effort is less likely to be wasted because each child's chance of survival is high. Again, she avoids having children during bad times and only conceives when circumstances are really favourable.

In all of these situations, stress is the contraceptive. But because women in different circumstances have different expectations, they are stressed by different levels of deprivation. Circumstances which would stress a woman in the suburbs of Western Europe or North America would not stress a woman living in a drainpipe in the Third World. It is a woman's *expectation* of living standard that determines the *number* of children she attempts to have. Whatever her expectation, as circumstances fluctuate and her expectations are exceeded or disappointed, stress comes and goes. As it does so, it influences just *when* she has each child as she strives to reach her preferred total.

Small families are not a new invention. For most of human history, from about one million years ago until as recently as ten to fifteen thousand years ago, all people lived as hunter-gatherers. Men hunted animals and women foraged for fruit and vegetables. Societies were made up of small, scattered bands of people. They had a good, protein-rich diet and most deaths were due to accident, predation and inter-group warfare rather than disease. The children of hunter-gatherers had an excellent chance of survival. Using nothing but the natural, stress-related methods we have discussed, women gave birth to only three or four children in their lifetime. Of these, two or three survived.

Large families did not appear until about ten thousand years or so ago, when agriculture brought a change of lifestyle. In the most fertile areas, large and concentrated communities developed, living on a carbohydrate-rich diet. Disease and infant mortality were rife. The average number of children was about seven or eight, but double figures were commonplace. Even so, whole families could be wiped out in days by virulent disease. As with the hunter-gatherers, on average, only two or three in each survived.

With the advent of 'modernisation', the infant mortality rate began to fall and so, too, a few decades later, did the birth rate. In Western Europe this decline in birth rate was again the product of natural family planning, for it preceded by up to a century the availability and use of modern contraception. So this reduction in family size to the levels found in modern industrial societies was due not to improved contraception technology, but to women subconsciously planning smaller families in response to the improved survival prospects of their children.

Many women will read this discussion of natural family planning with some cynicism, wondering how, if their bodies are so wonderful at family planning, they managed to conceive at precisely the time they least wanted. This is yet another example of the conflict between the conscious brain and the subconscious body.

As we have repeatedly seen, a woman's body may interpret its circumstances very differently from her brain. We encountered this conflict in Scene 15, in which the woman thought the last thing she wanted in the middle of wartime was to conceive. Nevertheless, her body actually went out of its way to become pregnant. In that instance, the body was proved right and the woman achieved most of her reproductive success through the son 'mistakenly' conceived on that occasion.

It would, of course, be wrong to conclude that the body is always right. No body is programmed to respond perfectly to every situation. The very fact that in every generation some people achieve a greater reproductive success than others indicates that some bodies make more mistakes than others. Quite possibly, though, bodies make fewer mistakes than brains. Yet recently, brains have been given greater power in this question of contraception. Modern contraceptives, such as the pill and the cap, at first sight would seem to take the control of conception away from the body and hand it much more firmly to the brain. But this change in control may not, as we have seen, influence the total number of children a woman may have. Rather, modern contraceptive methods supplement a woman's natural mechanisms by helping her to control when she has her children – and *with whom*, as we shall see in the next scene.

SCENE 17

How Forgetful

As the taxi pulled up, the woman heard her partner opening the front door. She listened as he came back into the house and picked up his bags. Holding her breath, she waited to see if he would come back into the room to say goodbye. But the front door closed, more loudly than usual, and he walked to the taxi without looking back. He would be gone for four weeks, and they were parting on bad terms.

They had been arguing on and off for a year now, and nearly always about the same thing. He wanted another child – preferably a son – but she didn't. She had been under pressure to come off the pill ever since the younger of their two daughters had started school. Now in her early thirties, the woman wanted to return to her career. Over the past few months, her partner had become increasingly disillusioned with his job and recently he had been ill. His disillusionment seemed to have stripped him of his health and energy and he had been off work for six weeks, returning only a month ago.

Those six weeks had been a nightmare. She knew that what he really needed to speed his recovery was peace and quiet but, in spite of herself, she just could not oblige. It irritated her, having him under her feet all day while she ferried the children to and from school, did the shopping and maintained the house. He had insisted on sitting around downstairs instead of keeping out of her way in bed. Try as she might, she could not avoid showing her irritation, and arguments were frequent. They had even argued over the new window-cleaner. He was a young man, earning money over the summer to help finance his place on some tropical expedition. Good-looking, confident and flirtatious, he had once commented on the bikini she was wearing in the garden. Her partner had overheard and accused her of making a pass at the young man.

Despite his return to work, their relationship had failed to improve. Early in the month she had vomited on and off for two days, and had been so worried in case she had lost the contraceptive protection of her pill that she had insisted on him using a condom. He had sulked all month. Then, last night, he had wanted unprotected sex – a gesture from her to him to see him through his four weeks away from her. She had refused, and they hadn't spoken since. Now he was gone, bad feeling unresolved. The day after he left, two things happened. She finished her current month's supply of the pill and she received an apologetic long-distance phone call from her partner. They made up over the phone as best they could. Then, over the next few days, they returned to their normal long-distance relationship, exchanging platitudes of concern and affection.

It happened just a week into his trip. It was mid-morning and she had just returned home from taking their two daughters to school and doing the shopping. Just as she had finished showering and drying her hair, the door-bell rang. She pulled on her bath robe and went to the door, to be confronted by her partner's immediate superior from work. He had called round to see if she had heard from her partner recently – because he hadn't. He had tried to phone her but had got no reply and, because there was an urgent message about a change of itinerary, he had decided to call round. She invited him in and, after they had discussed the reason for his call, politely offered him a coffee.

They had met socially a few times before. Once, much to the irritation of their respective partners, they had indulged in a flirtatious conversation at a drunken New Year's party. He reminded her of that night, and they laughed. Gradually, their conversation turned to the problems they were each having with their respective partners. She didn't notice what triggered her first flutter of sexual excitement as she busied herself with percolator and cups in the kitchen. All she knew was that at some point she became aware that only her hastily tied bath robe separated her body from his eyes and that he was watching her intently, waiting for glimpses of her nakedness. As the excitement took hold of her, she found excuses to touch his arms and back as she moved around in the kitchen. She laughed exaggeratedly at his comments and contrived reasons for bending and twisting in front of him, pretending not to notice as her robe loosened and the bulge in his trousers grew.

When the drinks were ready, they went into the sitting-room. She sat opposite him, breasts scarcely covered. Conversation became strained as they each waited for a sign that they would not be rebuffed. In a final move, she lifted her legs on to the chair, deftly pushing her bath robe between her legs to cover herself. Then, over the next few minutes, she watched his eyes flick backwards and forwards between her face and her genitals as she allowed her bath robe slowly to slip away, giving him an open display of her vulva. Minutes later they were having intercourse on the floor and ten minutes after that he had gone, suddenly embarrassed by what had just happened but telling her that her body had been irresistible.

This event began a week that in later life would seem to her like a scene from a film – a very bad film. Surely it couldn't have been *her* doing all of those things – acting out such clichéd events. At the time, though, she had simply seemed driven by some powerful force deep inside her. As the week progressed, she was to have many moments of fear – fear that somehow her partner would find out what she had done and fear that she had caught something. When, two days after this first infidelity, she realised she had forgotten to start her new pack of pills on time, she also feared that she might be pregnant. But for most of the week that had just begun, she was simply to feel a height of sexual restlessness and excitement that she hadn't experienced in years – and would never experience again.

Only an hour after her infidelity, she masturbated. She masturbated again the next morning. That afternoon, while sunbathing in the back garden, she heard the young window-cleaner arrive at the front. She was inside the house and up to her bedroom in seconds, standing in front of her bedroom mirror, back to the window. When she saw the reflection of the young man's face appear at the window, she slowly took off her bikini, then turned to face him. Feigning surprise at seeing him she waved unconcernedly, then busied herself with tidying the room, deliberately choosing jobs that involved much stretching and bending. When he rang the door-bell for his money, she opened the door naked, 'hiding' her body behind the frosted-glass door panel. She complained of the heat and, looking down at her nakedness, laughed that she had no money on her. Inviting him in while she went to fetch some, she joked that he could have sex with her instead if he liked.

No more windows were cleaned on her road that afternoon until, after her best climax during intercourse for years, she left to pick up the children from school.

She didn't masturbate again that week, but she had sex twice more with her partner's boss and once more with the young window-cleaner. That weekend, which she spent with her two daughters, it was as if she suddenly woke up from a dream. The excitement of the week disappeared, to be replaced by sheer disbelief at what she had done. On the Monday, she told both lovers that it had been fun but a mistake, and she was no longer available. The older man was relieved, the younger man disappointed.

When her partner returned, two weeks later, she told him about forgetting to start her pills on time but let him have unprotected sex anyway as a coming-home present. After that, in the days they were waiting for her period to start, she insisted he wore a condom.

She never did have her period. Three weeks after her partner's return, a pregnancy test confirmed that she was pregnant. Her partner was thrilled, even when the child turned out to be another girl. And he never did find out about his boss and the window-cleaner.

~

Most readers will recognise the seduction of the young window-cleaner as a cliché (as, later, did the main character in Scene 17 herself). It, or something similar, has been used as a not very imaginative 'dramatic device' in a multitude of films, plays and books. If the man involved is not a window-cleaner, he is an electrician, a plumber, a builder, a TV repair man or (in Britain, the biggest cliché of all) a milkman. In short, he is any man who has a legitimate reason for visiting a woman in her home while her partner is absent.

Indeed, so hackneyed is this scenario that there is a danger, if we are not careful, of missing the important point: namely, that the behaviour has become hackneyed precisely because it *is* so common. As such, it plays an important role in the promotion of sperm warfare – and hence in the paternity of children conceived via such warfare. Why should the woman in the scene, and so many like her, suddenly throw herself at two male visitors to her home? What is the significance of the saga of her contraceptive pills? And how did she benefit from her behaviour in terms of reproductive success?

When this scene began, the woman genuinely believed that she didn't want another child. In order to prevent conception, she was taking the pill and, when necessary, making doubly sure by forcing her partner to use a condom. Then, in the midst of what, sexually, was the most active week of her life, she 'forgot' to take contraceptive precautions. Was this lapse of memory really a mistake, or was it a subconscious strategy? And was it yet another example of a 'failure' destined actually to increase a woman's reproductive success? What this woman's body had really decided was that it didn't want another child by her partner. Her brain was then simply coerced into finding a convincing reason why they, as a couple, should avoid conception. That reason was so convincing, she even managed to persuade herself that she wanted to go back to work.

In her eyes, her partner had decreased in stature over the years. He had failed to live up to the potential he had shown when she committed her first two children to his genes. Now his health also was failing, and he was proving less than robust. As far as *she* was concerned, he had no saving graces in terms of intellect or character, either. In effect, her body had decided that a third child should be fathered by somebody with signs of a more robust genetic constitution than him. So when the opportunity presented itself, her body made sure that she took full advantage.

A woman is less likely to use contraception when she has sex with somebody other than her partner. And it is not always because the circumstances surrounding the infidelity make the use of contraception difficult. The first time the woman in Scene 17 had sex with her partner's boss or with the window-cleaner, it would have been difficult to insist on their wearing a condom. But on subsequent occasions she, and they, could have been better prepared – but they weren't. In any case, she wasn't dependent on these men and on condoms. Had she taken her pill at the appropriate time, she could still have avoided conception. But she forgot. Was it really an accident, a genuine lapse of memory? Or did her body subconsciously manipulate her into forgetting, so that she conceived to a man other than her partner?

The sudden but week-long surge of sexual excitement the woman experienced was the outward manifestation of hormones produced during her fertile period. We have already discussed (Scene 6) how women become more interested in infidelity at this time. Her sudden loss of sexual interest at the end of her week of infidelity marked the end of her fertile period and, as it happened, the beginning of her pregnancy. By the time her partner returned and accepted an unprotected intercourse as a coming-home present, the woman will already have been two weeks pregnant.

Women, like most female animals that hide the fertile phase of their menstrual cycle, continue to have intercourse well into pregnancy. This is the final touch by which they can confuse the males around them. If a woman lost interest in sex as soon as she conceived, it would be a clear signal to the males around her that conception had occurred. This would allow each of them to make some assessment of who could and could not be the genetic father. Continuing intercourse well into pregnancy guarantees the ultimate confusion of all potential fathers. This explains why our main character was keen to have unprotected intercourse with her partner on his return. It meant that in his eyes, and even in hers, he *could* have been the father of her third child, even though he wasn't.

Given such a short-lived and ideal opportunity for undetected infidelity, the woman's body made a shrewd move in collecting sperm from more than one man. She gained two benefits. First, she halved her chances of being unlucky enough to collect sperm from a man who just happened to be infertile (10 per cent of men, largely due to sexually transmitted disease – Scene 11), despite appearing to be a suitable genetic father (Scene 18). But secondly, by putting two men's armies into competition, she increased her chances of being fertilised by an ejaculate competent at sperm warfare (Scenes 6 and 21). She might never again have such a perfect opportunity to conceive a child with better genes than her partner could provide – not without leaving him, anyway. We do not know which of the two competitors was actually the father, but whichever he was, he was the man who won the sperm war she had promoted.

When a woman has sex with two men over a short period of time, she has three ways of influencing which of them fathers any child that may result. First, she can have sex with one of the men at a more fertile phase of her menstrual cycle (Scene 6). Secondly, she can retain a larger sperm army from one of the males (Scenes 22 to 26). Thirdly, like the woman in Scene 17, she can use modern contraceptive techniques.

If a woman uses a barrier method, such as cap or condom, with one man but not the other, she can prevent the former from entering sperm into warfare at all. Alternatively, by using the pill and then not using the pill, like the woman in the scene, she can influence which man's sperm is most likely to have access to an egg. Indeed, our main character made full use of modern contraception in ensuring that she did not conceive via her partner. The other two men, however, were given equal chances. All they had to do was win a sperm war.

As we discussed in Scene 16, modern contraception may not have much influence on *how many* children a woman has in her lifetime. But it has provided her with a powerful and efficient tool with which to enhance her natural ability to time *when* and *via whom* she conceives. Rarely is contraception technology used in this way consciously. But in the hands of a woman's urges it is a powerful new weapon by which she can increase her reproductive success.

7

Shopping Around for Genes

SCENE 18

Spoilt for Choice

Despite the open windows, the room was hot and humid. The naked woman moved slowly off the bed so as not to wake the young man sprawled next to her. Without dressing, she walked downstairs, through the open french windows, and into the bright sunlight. The tiles of the patio were hot to her feet. She paused for a second, then ran a few steps and dived into the large swimming pool. The blissfully cool water washed the remnants of her recent intercourse from her thighs and pubic hair and, as she swam the length of the pool, carried away her flowback.

She had been in the water about ten minutes when the young man she had left on her bed appeared on the patio. She watched his naked, tanned and muscular body run across the hot tiles and dive not quite perfectly into the water. He swam strongly over to her, trod water by her side, then kissed her. After a few minutes together, enjoying the water, she asked him to go. Without argument, he swam back across the pool, pulled himself out of the water, and disappeared into the house to shower and dress.

The woman also left the pool. She dried herself with a towel hanging from the pool-side lounger, then walked down the steps on to the lawn, relishing the coolness of the grass beneath her feet. There was no danger of her nakedness being seen. The garden was so large, the bordering trees and security fencing so high and the neighbours so far away, that her privacy was complete. When the young man appeared at the french windows to shout his farewell, she was on her way back to the patio. She felt like a Greek statue as she stood in the middle of the lawn and waved goodbye. For a woman only months from her fortieth birthday and with three children, one now in her mid-twenties, her body was in impressive shape.

As the young man turned to leave, she walked back to the patio. In five years, he was probably her best find yet. Her advertisement was simple but effective: 'Part-time gardener required to maintain large garden over summer. Ideal for student during summer vacation'. Each year she had interviewed over twenty applicants. Her choice was based on the student's looks, intelligence, maturity, self-confidence and sexual aura. She took it as testimony to her own good looks and judgement that it had never taken her more than a fortnight to seduce any of her choices. The result was that for two days a week, over about three months each summer, she had the sexual companionship of a young man nearly twenty years her junior. And she had her lawn mowed and her garden weeded.

This year she had hired two men, both medical students, whom she had found equally attractive. One worked on Tuesdays, the other Fridays. As she stretched out in the sun on the lounger, she toyed once again with the fantasy of having both young men in bed with her at the same time. But that was for the future. As she drifted in and out of a sun-drenched sleep, she reflected on the events that had given her such a good life.

It all began on her fourteenth birthday. She had been out with her mother, shopping around for a new pair of jeans. It was a freezing cold winter's day, and while they were shopping it had started to snow. The town was a forty-five-minute bus ride away from their home, even on a good day. By the time they arrived at the bus stop, the snow was settling fast. The bus never came. But the man in the large and very expensive car did.

He wasn't really a stranger. When he wasn't at his apartment in the city, he lived in a virtual mansion in their home village. Her father worked for him as a gardener cum handyman, helping to maintain his house and its grounds. Although for years she had seen him driving around, she had never really met him before. All she knew was that he was about fifty years old, very rich, and apparently had no children, despite having lived with the same woman for about twenty years. She had heard her parents talk about infidelity and children by other women. She had also heard them talk about infertility and women's problems, but had taken little interest.

On the journey home she had been struck by his friendliness. At fourteen, she was well developed and already attractive, with a maturity and self-confidence that would have graced a twenty-year-old. During the journey, she found herself doing most of the talking. She liked and felt comfortable with this man and had no hesitation, a week later, in accepting another lift from him when he drove past her at the school bus stop. After that, he drove her home with increasing frequency. Her schoolfriends teased her a bit about it, but it didn't bother her.

The following summer, during the long vacation, she lost her virginity. Her lover was a boy of seventeen whom she and her friends had worshipped from a distance for months. After the trauma of their first intercourse, she began to enjoy their sexual activities. She became the envy of her friends as she described in great, though sometimes imaginary, detail her sexual adventures with this young idol.

That autumn, she resumed her journeys home from school with her father's employer. She was in his car the day before his partner was diagnosed as having terminal cancer. After that, she didn't see him for several months as he and his partner moved to the city for her to be treated. He did not reappear until a few weeks after his partner's death which, she discovered later, had been the day after her fifteenth birthday. From then on, whenever he wasn't in the city on business, he made a point of picking her up from school. Soon, they actually discussed which days he could do so, and she would wait for him.

Winter cut down on the sexual activity with her boyfriend until he passed his driving test and acquired a virtually derelict car. Then, they became adept at cramped sex on the cold and badly sprung back seat. She began to fantasise about doing the same with her father's employer in the luxury of *his* car. It would be like being in bed. Now, stretched out in the sun by the side of her swimming pool, the woman still remembered clearly the move that changed her life. On their way home from school, waiting at traffic lights, she had put her hand on his thigh and leaned across to kiss him on the cheek.

They never did have sex in his car. But within a week of that kiss she was in his bed, experiencing for the first time the difference between sex with a man of fifty and sex with a boy of seventeen. Throughout that spring and summer she had sex at least twice a week, more or less alternately with her boyfriend and the older man. Neither knew of the other's existence. It was autumn before she and her mother realised she was three months pregnant.

She refused to discuss paternity with anyone except her parents and her father's employer. Her parents were told that the father was the young boyfriend who had just left for college and whose parents had now moved out of the area. Her father's employer was told that the baby was his, as well it might have been. He feared conviction for having sex with a fifteen-year-old girl. But he also felt genuine affection for her and for the child he believed was his. Explaining to her parents that he would do the same for any employee who was in trouble, he offered to help with the child's upkeep. On the strength of this extra income, her parents volunteered to look after the child, a girl, while their daughter finished her education.

When, despite the distractions of motherhood, she performed well in her examinations, the man again offered to help. This time he provided money for her to have a college education. During her two and a bit years at college, she had about ten different sexual partners. Even so, twice a term she would arrange to meet her benefactor at his city apartment for a weekend of sex and a taste of the high life.

She never did graduate. In her third year, faced with a choice between months of hard revision and a tempting offer from her benefactor, she left college. They travelled extensively for a few months and then, on their return, began to live together with the daughter, now aged six. At her urging, he sold his mansion and bought the wonderful house in which she now lived. They spent nearly ten years together in comfort and luxury, travelling the world, and mixing with equally wealthy people. They also had a further two children, both boys.

All three of the woman's children were raised by nannies, sent to boarding school, and spent very little time at home. Her partner was without doubt the father of her elder son, but she could not be sure who was the father of the younger one. It *could* have been her partner, but it could equally have been the politician with whom she had sex every day for a week at about the relevant time. Moreover, if he had been a month later taking her to bed, so too might have been the surgeon, a family friend who had treated her predecessor for cancer.

Her partner lived just long enough to see the younger son's eighth birthday. Then, aged sixty-five, he died of a heart attack. That was ten years ago. She inherited the house and more than enough money to live very comfortably – she continued to pay for her children's education, travelled when she wanted and indulged in occasional luxuries, like young gardeners. After her partner's death, there had been no shortage of men eager to share her life. Many were widowed or separated and most were rich. She was rarely without a sexual partner, and often successive partners overlapped. However, she stubbornly refused to allow anybody to live with her permanently. Besides, she was increasingly attracted to struggling young men, full of drive and ambition, rather than to men saddled with the complacency of success or inheritance.

Her daughter, now aged twenty-five and living abroad with her partner, had just announced her first pregnancy. Her sons, aged eighteen and nineteen, were both at college studying medicine. She took great pride in all her children, particularly her sons – not because of what she had done for them, which in truth was very little, but for what they were and for what they had done for themselves. The boys were very different from each other, perhaps reflecting a different paternity. But they were both good-looking and intelligent, precociously mature and confident, yet still gentle and caring. She had no doubt they were going to break many a girl's heart.

Earlier, in their mid-teens, they had been in the habit of bringing home friends from school to stay for a while during the vacations. Many a young boy had been shocked and embarrassed at the family's practice of swimming naked together in the pool. In two weeks' time, her sons would begin their summer vacation. This time they would stay with her for only a week before travelling abroad, and, moreover, they were bringing girls with them, not boys. Mischievously, the woman wondered if they would swim naked with her this summer – and, if they did, whether their girlfriends would join them.

She got up off the lounger and sauntered across the patio, dragging her towel behind her. It was time to shower and dress so as to be ready for her evening's escort. As she made her way indoors, she enjoyed yet again the thought that for a gardener's daughter she had done rather well.

~

In congratulating herself on her life, the woman in this scene was of course measuring her success on a hedonistic scale. By her own and probably most other people's yardsticks, she had indeed done 'rather well'. And even on a biological scale, she had done rather well.

Few factors have more influence on a person's reproductive success than their selection of a mate or mates. Yet mate selection is complex, particularly for a woman. More often than not she has to compromise in many different ways. The scene we have just witnessed is the first of four in which we shall explore two matters – the problems that people, particularly women, face when selecting a mate; and the methods they employ in solving those problems.

Here the central character successfully cleared all of the obstacles to reproductive success that women normally encounter when choosing a mate. First, through her choice of long-term partner, she engineered an environment conducive to the easy and successful raising of children (from the point of view of being in a position to offer them every opportunity, anyway). Secondly, she managed to collect some of the most sought-after male genes in her vicinity. As a result, she produced children with the looks and ability to make the most of the comfortable environment into which they were born. Her strategy was risky, but innate ability was on her side. She made the most of her daring and cunning, her composure and good looks, and successfully walked the tightrope of disease, discovery and desertion.

In choosing a man or men with whom to share her life, a woman has two major issues to consider. On the one hand, she needs a man who can help her raise her children. On the other, she needs genes that in combination with her own will produce attractive, fertile and successful children. The better the environment and the better the assistance, the more fully each child will achieve his or her genetic potential.

A woman's difficulty is that she has a much wider choice of men to provide her with genes than she has of long-term partners. She could probably persuade many men of her choice to give her their genes – it takes only a few minutes of sex, after all. Her options for a long-term partner, though, are much more limited. Most men in most societies have not the time, the energy nor the resources to help support more than one woman and her children at any one time. Her choice of long-term partner is therefore restricted to those men who are unattached, ready to desert their current partner, or who have so much time, energy or wealth that they can support more than one family.

An equally difficult problem is to identify which of the few available men would make the best long-term partner. The most reliable way would be to look at past performance, but inevitably the best long-term partners are already paired to other women. Much of the woman's choice, therefore, is limited to young and unattached men who have not yet proved themselves as long-term partners. All she can do is look for signs of potential, and hope that her judgement is accurate.

Surveys of many cultures around the world consistently show that, in looking for a long-term partner, women prefer men who have, or have the potential of, wealth, status, stability and durability. In the past, in all cultures, the children of women paired to men at the top of the scale for these qualities had a far greater chance of survival, health and subsequent fecundity. The same holds true even in today's industrialised societies.

The preferences are clear, but for most women a level of compromise is necessary. One man may be wealthy but uncaring; another may be of high status but unstable; yet another may be poor, but stable and caring. So, inevitably, a woman has to opt for the best compromise. Of course, she does not have to stay with her first partner. Again, studies show that when a woman leaves one partner for another, she invariably moves up the scale to a better compromise.

In choosing a man to help raise her children, a woman is only secondarily impressed by looks, whereas in choosing a short-term partner for sex, looks are much more important. The features she finds most attractive are clear eyes, healthy skin and hair, firm buttocks, a waist that is about the same in circumference as his hips, shapely legs, broad shoulders, quick wit and intelligence. She is also attracted by symmetry in his physical features. These various qualities are all reasonably reliable indicators of genetic health, fertility and competitiveness. As such, they imply a genetic constitution that would also be desirable in her children.

Since women are seeking different attributes in short-term and long-term partners, but have more choice of short-term, they may again have to compromise. They have two main options. They can choose the best available long-term partner, and then rely on infidelity to obtain the best genes. This *can* succeed, but only if they successfully avoid the disadvantages of infidelity that we have already discussed (Scenes 8 to 11). Alternatively, they can choose a man who, although neither the best provider of genes nor the best partner, is at least the best available compromise.

As we have found in so many aspects of reproductive success, women's behaviour and experience are mirrored by other animals. In this instance, women are not the only females who have to trade off between male support and male genes. One of the most revealing studies of the problem concerns a bird, the blue tit. The females of this species show all of the behaviour we have just described for women. Those lucky ones paired to genetically superior males with the best territories are totally faithful. Neighbouring females, paired to genetically inferior males, take every opportunity to seek infidelity with the superior males. They sneak into the better males' territories, solicit intercourse, then return unobserved to the partner they have just cheated. On average, about a third of young birds in a nest have not been sired by their mother's partner. Actual levels range from 0 per cent in the nests of the most favoured males to about 80 per cent in the nests of the least favoured ones.

A surprisingly similar pattern is found in humans. On average, about 10 per cent of children are not sired by their supposed father. Some men, however, have a higher chance of being deceived in this way than others – and it is those of low wealth and status who fare worst. Actual figures range from 1 per cent in high-status areas of Switzerland and the USA, through 5–6 per cent for moderate-status males in Britain and the USA, to 10–30 per cent for lower-status males in Britain, France and the USA. Moreover, the men most likely to sexually hoodwink the lower-status males are men of higher status. Anthropological studies have shown precisely the same pattern. Men of higher wealth and status obtain partners earlier, start to reproduce earlier, are less likely to have their partners impregnated by other men and are more likely to do exactly that to other males. So in all ways, men of wealth and status have the potential to be reproductively more successful than their lower-status contemporaries.

Returning to birds – it is rare, unless the male is infertile, for all of the young in 'his' nest to have been sired by other males. It is as if the female always gives her partner *some* chance of paternity so as to retain his help in paternal care. The same may be true for humans. Moreover, when we look at which children are most likely to belong to their partner, women show a clear pattern. As with the woman in Scene 18, the child most likely to have been sired by a woman's partner is the second; the children least likely are the first, and particularly the last. The reasons, though, are slightly different for first and last.

Often, a woman is already pregnant when she settles down with a long-term partner, and occasionally this partner is not the father of her child. Sometimes he knows this and takes on the woman and her child anyway (Scene 15), for reasons we have discussed (Scene 9), but sometimes he doesn't know. The woman is *least* likely to be unfaithful in the weeks or months preceding the conception of her *second* child. Subsequent children, however, are more and more likely to be the product of infidelity.

Identifying the best partners and the best providers of genes, then pursuing the best available compromise, is only one aspect of a woman's problem in obtaining a mate. Having selected him, she then has to recruit him into the chosen role. She can do this only if the man finds her attractive enough. If she cannot attract her first choice, she then has to compromise yet again. The man she finally recruits will be the best compromise among those she wants and those she can attract. The woman in Scene 18 was successful because a rich and successful man, thirty-five years older than herself, picked her out from the crowd. Then, without her infidelity being detected, she set about collecting genes from other successful men. She succeeded once again because these men considered sex with her to be worth the risk. In short, she succeeded because she was attractive to men, both as a partner and as a lover.

So, what was it about this woman that men found so attractive? What criteria do men use in choosing partners and lovers, and how do these criteria differ from those used by women?

Basically, men select women for their health, fertility and fidelity – though not consciously, of course. On seeing a woman, men do not immediately remark on her potential for bearing and raising children. Nevertheless, the features that men's bodies are programmed to find attractive are precisely those that do reflect this potential. Unlike a woman, a man uses similar criteria whether he is selecting a partner or a lover – for both, his primary concern is with looks and behaviour. An important feature is body shape, particularly the ratio of waist to hip. Irrespective of whether a woman is thin or fat, men prefer someone whose waist measurement is about 70 per cent of her hip measurement. This preference is remarkably constant both throughout history (to judge from statues, paintings and 'girlie' magazines) and from culture to culture (to judge from rock-paintings and figurines). In some cultures, men prefer thin women; in others, fat. But in all cultures, they prefer women with waists significantly narrower than their buttocks. The explanation is that this shape reflects a good hormone balance, good resistance to disease and good fertility.

In addition to shape, men all over the world also respond strongly to clear eyes, healthy hair and skin, and the shape of the face, particularly its symmetry. Again, these features are strong indicators of health and hence fertility. Men of most cultures also respond to breast size and shape, though actual preferences vary and there is no simple link between the appearance of a woman's breasts and her ability to lactate and sustain a child. Finally, men respond strongly to certain character traits, such as meekness and dependence, that might indicate potential fidelity. Such traits, though, are relatively easy to fake, at least for short periods.

There is another difference in mate preference between men and women. As long-term partners, women tend to prefer men who are older than themselves. Such men have had more time to prove themselves and more chance to amass the resources a woman will need to sustain any children she may have. Unless he is very wealthy, however, a man older than about fifty becomes less and less attractive to a younger, still fertile, woman – because of the increasing risk that he might die before any children she has with him become independent.

Men, on the other hand, prefer females who are old enough to be fertile, but still have most of their reproductive lives ahead of them. This way, they get more children from their investment. Whether a man is twenty or seventy, therefore, his preferred age for a new partner is about twenty, or even younger. Hence the frequency with which the most successful of men leave their middle-aged partner and their first family to set up home and begin a second family with a much younger woman.

An older but still fertile woman might also gain from a liaison with a younger man. This is because she can judge a man's physical quality more easily when he is at his physical peak. But more often than not such a man is in no position to maintain an older woman, let alone her existing children and any new children their liaison might produce. Consequently, older women are most likely to target young men for acts of infidelity from within a secure long-term partnership, and are less likely to choose them as long-term partners.

The woman in Scene 18 had been freed from constraints by her personal wealth. She could have invited any of her 'gardeners' to live with her for a while and perhaps give her another child – they had all been carefully chosen for their genetic potential. She could undoubtedly have raised another child by such a man without prejudicing either her own health or the success of her existing children and future grandchildren. Nor need she have short-changed her new child. By the time we left the scene, the woman had not followed this course, but she still had time to do so. Many rich women do.

Finally, once a woman is post-reproductive or once a man has ceased to be able to attract younger, fertile women (Scene 11), their criteria for mate choice change. Choice of a long-term partner can still influence their reproductive success, but now of course it is not via any children that they might produce together. It is instead via its effect on the survival and reproduction of any children they each may already have from a previous relationship (Scene 11). Wealth, status and the potential to be a good step-parent or stepgrandparent now become of primary importance to both sexes in their choice of a partner. All the same, most people still have to compromise.

We can see quite clearly that nowhere is the conflict of interests between males and females greater than in the selection and recruitment of a partner – at any age. Everybody is seeking the best partner in their preferred category, but may not themselves match up to that chosen person's preferences. Competition is fierce. Everything is compromise, and time is limited. If a person settles too readily for a poor compromise, they may miss the chance of a much better compromise later. However, spending too long in search of the best compromise can be equally disadvantageous. He or she may then pay the price of having to settle for a worse compromise, or even of failing to attract anybody at all. The best prizes go to the people who judge correctly when to continue their search and when to settle for what they can get – if only for the time being.

There is a fascinating consequence of the criteria women use for mate selection – particularly their preference for, and greater fidelity to, men of wealth and status. The sons of such men achieve greater reproductive success than their lower-status contemporaries. They do this not only through their long-term relationships, but also because they have the same above-average opportunities to sire children with the partners of other men as did their fathers. It follows from this that a woman paired to such a man will achieve greater reproductive success if she produces sons than if she produces daughters.

The greatest number of children ever claimed by a man is 888 (by an ex-emperor of Morocco); that by a woman, sixty-nine (over twenty-seven pregnancies). And even at a more mundane level it is much easier for a man, potentially, to have more children than a woman – for the obvious reason that whereas a man can have his children with several different women, a woman (until the recent advent of surrogate motherhood) has had to have all her children herself. A *successful son*, therefore, can give a woman far more grandchildren than can a successful daughter. The higher a son's eventual social status, the more successful he is likely to be. Since wealth and status, as well as the genetic potential to achieve wealth and status, can be inherited, we might expect higher-status couples to produce more sons than lower-status couples – and they do. Studies around the world have shown a male-biased sex ratio among the children of couples of higher status (take, as an example, the children of the men and women listed in the national *Who's Who*). Usually the bias is statistical rather than obvious – about 115 boys for every 100 girls – but sometimes it can be impressive. The presidents of the United States, for example, have between them produced ninety sons and sixty-one daughters, the equivalent of 148 sons for every 100 daughters.

So why don't all women produce an excess of sons? Actually, to some extent they do. On average, about 106 boys are born for every 100 girls. But because boys are more likely to die during childhood, by the time the survivors start to reproduce the proportion is about equal. Even so, not only is the average woman less likely to have a boy than a woman paired to a high-status male, but women paired to low-status males and women without a partner at all are more likely to produce a daughter. Why?

The answer is that a son is a much more precarious reproductive option than a daughter. Despite his *potential* to produce large numbers, he is more likely to die before he begins to reproduce, and he has much more chance of not reproducing at all, even when he tries. If they are not reproductively competitive, sons are a poor option. Imagine a society in which all women produce only two children: for every man who sires six children with three different women, there will be two who fail to sire any. There are two ways in which a daughter is the safer option: first, although relatively few daughters produce large numbers of grandchildren for their mothers, relatively few fail to produce any; second, a mother can be certain that all of the grandchildren produced through a daughter are hers. She cannot be so certain about the grandchildren apparently produced by a son.

Thus, only when there is a very good chance that a son will not only survive but will also be reproductively competitive against other males is it worth producing a boy. So, in principle, we should not be surprised to find that women without long-term partners and women paired to lower-status males produce an excess of daughters – or to find that those paired to higher-status males produce an excess of sons. Nor should we be surprised that, in between these two extremes, most women compromise and show no bias towards either sex.

The woman in Scene 18 got it exactly right when she produced a daughter from her 'accident' while she was still at school, and two sons when she was paired to a wealthy man of high status. Just *how* she will have achieved this bias is not known. The explanation is not that higher-status males produce an excess of 'male' sperm – the casual lovers of such men produce an excess of daughters, as they should. Nor does it seem likely that high-status males introduce an excess of 'male' sperm into their partners but an excess of 'female' sperm into their lovers. The only reasonable explanation is that the bias is generated by the woman. Either her body biases the proportion of male and female sperm allowed through to the fertilisation zone in her oviducts, or she is selective about which embryos she allows to implant following fertilisation. Maybe, if the embryo is the 'wrong' sex for her circumstances, her womb does not let it implant (Scene 16).

To most people, the whole process of mate selection is a minefield of distractions and pitfalls, particularly during late adolescence as a boy or girl searches for their first long-term partner. The woman in our scene subconsciously wove her way through this minefield with precision. She did so by virtue of the excellent genes she herself had received from her parents. As a result, she ended up with three children, and perhaps eventually more, all sired by a man or men who were reproductively outstanding in their peer group. Her daughter had already begun to produce grandchildren; her sons had great potential, not only to produce their own families but also to be the targets of other women's infidelities. She had every indication that her descendants would multiply and flourish. In later generations, her genes would mark more of the population than would the genes of her less successful contemporaries. Her life had been a success – both biologically and hedonistically.

SCENE 19

Fair Exchange

It was Saturday night and the two couples were sitting cross-legged in a circle on the floor. With plenty to drink by their sides, they began to play cards. Each of them was privately tingling with nervous and sexual excitement. They had great hopes that the next hour would be a sexual landmark in all of their lives.

They had known each other for seven years. Although they had always found each other's partner sexually attractive, no infidelity had ever occurred. The first couple, despite five years of unprotected sex, had no children. The second couple had two. In recent years, for different reasons, each couple's relationship had been foundering.

The childless couple felt an undeniable emptiness in their lives, which they could not shake off despite all their money and travelling. When they first met, they had seemed perfect for each other. He was tall, muscular, ambitious, witty and dominating. She was effervescent, open and liberated, always dressing provocatively. Women surrounded him wherever he went and he could have had a new sex partner every week if he had wished – and he did . . . often. She, also, was much sought after. Once regarded as an item, they gained great kudos from being each other's partner and, despite their earlier promiscuities, had settled into a serious relationship.

Four years later, after a year of unprotected sex, they had been screened for infertility but there was no clear indication that either was infertile. She was ovulating, had no blocked tubes and seemed normal. He was producing rather large numbers of sperm, but otherwise also seemed normal.

At first, they responded to their problem with intense sexual activity, but after a further year of failure they gradually lost interest in routine sex. Secretly, they both hoped their combined failure was the other's fault, and this thought had led them to be critical of each other's sexuality. She now saw his size and power as futile, boorish and selfish. He now saw her free conversation and provocative dress sense as public vulgarity, a front to hide her private frigidity.

The couple with children were very different. Both were attractive in their own way, but were much less extrovert than their friends. What little social kudos they had came from their association with the childless couple. The man was small, quiet, hardworking and reliable. The woman was demure, sympathetic and maternal. It had taken them eight months of effort to conceive each of their two children but eventually it had happened and, as a parental team, they were wonderful. When he was not working they were hardly ever out of each other's company.

Lately, however, increasing financial hardship had produced friction between them. Their situation had been aggravated when his employers, dismayed by his lack of ambition and charisma, moved him sideways to a lower-profile position. Now, increasingly frustrated by her partner's mundane job, she was often critical of his inability to improve their situation. For his part, having pushed his partner into full-time motherhood, he now found her demure quietness dull and unattractive.

Recently, each couple had sought to bolster their flagging sexual interest with pornography. But over the weeks the impact of the images had worn off. Then, when the four of them started to watch films together, excitement had been briefly rekindled. Even that source of excitement, though, had eventually begun to wane – until tonight. The idea for the 'game' they were about to play had been seeded by a film showing partner-swapping. In the video they had just watched, the men had thrown their car keys into a pile in the centre of the room. The women chose a key, then paired off with the owner. Soon the room was full of naked, copulating bodies.

It was the childless woman who with typical brashness had suggested they should try swapping. At first they all treated the suggestion as a joke, but as they became increasingly drunk they began to talk about the possibility seriously. Soon they were discussing the practicalities of the exercise rather than its desirability. They devised a card game, based on chance, not strategy, that required the winner to remove an item of clothing. It was decreed that the man who was the first to lose all his clothes should have sex with the woman who lost all hers first.

A run of wins left the smaller of the two men naked almost before the others had started. He was far from muscular and had relatively tiny testes, but his penis was larger than average. As he settled back to watch the two women progressively expose more and more of their femininity, he became increasingly proud of his erection. In the event, the cards were disappointing. In a late run, it was his partner who was next to be naked. As she removed her knickers, she both surprised and excited the other couple by revealing a triangle of pubic hair that struck them as incongruously large for such a demure woman. An immediate suggestion that the rules should be changed so that partners did not have sex with each other was defeated by the alternative idea that they should still have sex, but in full view of the other couple.

Less exhibitionist than their friends, the 'lucky' couple found it difficult to begin their performance. Laughter, self-consciousness and a rapidly dwindling erection almost ended the enterprise. Eventually, as they lay together, it was the serious stroking and kissing of their bodies by the other couple that took them across the threshold from embarrassment to intense sexual arousal. The intercourse they had, while still being caressed by the other couple, was the most exciting either had ever experienced.

Aroused almost to the point of spontaneous ejaculation by what he had just seen, the childless man suggested they should abandon the cards. He should now have sex with the already naked and prostrate woman rather than with his own partner. Almost before anybody could comment, he was on top of her. The woman was to relive this moment many times over the next few years. The emotion she was always to remember as she felt him, penis even larger than her partner's, slip gently into her was neither guilt nor excitement. It was instead an overwhelming sense of welcome. Excitement followed but was short-lived. He had been so aroused that he ejaculated within seconds.

All now naked, there followed a crude and drunken interlude of about half an hour while they first waited for, then tried to help, the first man to recover his erection. The childless woman, who had perhaps wanted all this to happen more than any of them, began to fret that she would not have her turn. The pressure on the first man was intense but eventually the prospect of sex with a different woman, combined with the now totally abandoned sexual encouragement he received from both women, did the trick. At the childless woman's suggestion, he entered her from behind as she knelt down, bending forward over a chair. Watching his partner have sex with another man excited the childless man once more. He could scarcely wait for his friend to withdraw before taking over.

They never repeated the exercise. A month later, both women discovered they were pregnant. They both assumed that the father of their children was the one who had already proved his fertility. But they were quite wrong. The father of both was the one who had previously seemed infertile.

The couples were to stay together for another five years. For a while, until after their babies were born, both relationships stabilised. As the children grew, however, both relationships once more deteriorated. Despite continuing to have unprotected sex, the extrovert couple failed to produce another child and their mutual recriminations intensified. The man began an affair with the other woman. This time, they kept it secret. Eventually, though, they confessed their infidelity, both couples separated, and the unfaithful pair began to live together, along with the woman's three children.

Financially, the more extrovert man's life went from success to success and he was able to maintain in some comfort not only both of the children conceived that night but also his two stepchildren. By now he was convinced that he was infertile. However, his increasing wealth and status and his continuing physical attractiveness made him even more of a magnet to other women. He was often unfaithful. His partner knew about his infidelities – but she agreed to tolerate his behaviour in exchange for a guarantee that, no matter what happened, she and her three children would be financially secure. In the years to come, he unknowingly sired two more children via his promiscuity.

~

In previous scenes, we have watched men and women going to great lengths to hide their infidelity. If they failed, they suffered. Yet here we meet four people who are openly unfaithful. Partner-swapping is not particularly common in Western society, but it occurs often enough to form a recognisable part of the rich mosaic of human sexuality – a part, moreover, that promotes sperm warfare. A survey in the USA in the 1970s revealed that almost 5 per cent of couples had, at one time or another, openly swapped partners with another couple, though nearly 80 per cent of these had done so only once or twice. What circumstances can lead such behaviour to actually improve a person's reproductive success?

The answer lies in the way criteria for choosing partners and lovers can change with time during the course of a person's life. Of course, when open infidelity was first mooted in Scene 19, the conscious preoccupation of the four characters will have been with superficial matters. Excitement and fear will have been relished in equal proportions – the excitement of imminent infidelity will have been tempered by fears concerning their bodies and their sexual performances. Subconsciously, however, their bodies will have been wrestling with more important matters. Would the reproductive benefits of their planned actions outweigh the reproductive costs?

The more extrovert couple had an easy decision. After five years of unprotected sex, they had failed to produce children. Each thought their partner might be sterile, and here was a golden opportunity to have sex with someone of proven fertility. Both their conscious and their unconscious brains might well have reached the same conclusion.

The other couple had a more difficult decision, though consciously neither would probably have appreciated their body's logic. Their family was rapidly outgrowing their home and finances. Even two children might be too many for them to raise successfully without an improvement in circumstances. So how could they each have another? The man could manage to have a third child only if he could find somebody who would raise that child without his help. His friend's partner offered just such a possibility. His own partner, on the other hand, could have a successful third child only with better financial support than he, her present partner, could provide. Her friend's partner had potential, argued the woman's body. Not only was he wealthy, he was in a relationship that was looking increasingly unstable. Maybe, if she had a child which he thought might be his, he would help her financially. Or she might even be able to wrest him from her friend, his current partner. On top of all that, her body could see a benefit in conceiving via this man: not only was he very attractive to women, he had also achieved a higher status in life. Genetically, a child by this man could be far more successful than another child by her current partner.

Of course, some of the potential costs of infidelity were still there for both couples – risk of disease, for example. But most of the dangers usually associated with infidelity were minimised by the nature of the situation. All of the characters knew and agreed to what was going to happen, so there was no need for deception and hence no risk of discovery. By agreeing to have sex in view of each other, they removed all possibility of deceit. Both relationships were already unstable, and the risk of anybody being deserted by their partner because of the night's events would have seemed no greater than before.

All four, therefore, subconsciously computed that they might gain from a bout of partner-swapping. And as it turned out, all four were correct. This is easy to see for three of them. The childless woman conceived. Her partner sired two children in one night. The other woman struggled initially, but eventually wrested the first man from his partner, benefited from his wealth, and successfully raised three children. Without that night of abandonment, she might have raised only two – and those, with difficulty. Only for the other man is it less easy to see the value of his decision. Of course, he *thought* he sired two more children that night. But he didn't. Moreover, for five years he had the stress of raising a third child (which wasn't his, though he thought it was). Eventually, he was deserted by his partner for a man with whom she had first been unfaithful with his consent. As a result, he also lost day-to-day contact with the children, for whom he cared very much.

At first sight, his may not seem much of a success story. However, as a result of that night's events, the two children who really were his were raised more easily and given a better launch into life than he could ever have given them. He still saw them often and gave them what help he could. From time to time they lived with him for short periods, and they probably never really felt abandoned by him. If he hadn't agreed to that night's events, he might have struggled to raise his two children. And it is worth bearing in mind that, as we saw in Scenes 9, 11 and 16, such a struggle can sometimes have dire consequences. He might still eventually have been deserted by his partner, but in less favourable circumstances. In particular, he might well have been less fortunate in the man who took his place in raising his children. Indeed, a key factor in the less extrovert man's success was that his friend did turn out to be a good stepfather. Of course, in part this may have been because he kept a close eye on the way his children were being treated in their new situation. Mostly, however, it was because his ex-partner tolerated her new partner's infidelities. The latter was not by nature monogamous, but as long as he could be promiscuous he was happy to help support her children. In effect, she traded his sexual freedom for a guarantee that, no matter what happened, she and her three children would be financially protected.

All four gained from their night of sexual abandonment. But if things had turned out differently, and the previously childless man had failed to win those two sperm battles, he would have lost out in a big way. But, in fact, he was most unlikely to lose. He was a sperm wars specialist, producing too many sperm. His friend was not, producing too few. Put the specialist in a monogamous situation and he was sub-fertile. After each insemination, huge numbers of his killers and egg-getters clustered around his partner's egg. Multiple sperm entered the egg simultaneously and dense concentrations of deadly chemicals were released by the surrounding sperm hordes. The egg always died. But send his sperm in to battle, and he was virtually invincible. His huge armies would decimate those of lesser opponents and, having been reduced in number themselves, would then send the correct number of sperm on to seek an egg.

On that night of sexual revelry, the less extrovert man produced small armies, rich in blockers (Scene 7) and family-planning sperm (Scene 16). His reproductive strategy in the past had centred on guarding his partner from other men. Then he could use long-term, routine sex to produce children at long but measured intervals when times were good (Scene 16). Twice, he had needed eight months of unprotected sex to impregnate his partner. Even so, he would probably have produced more children with his partner had he been more successful socially and financially. But in sperm warfare he was doomed to failure.

We discuss the existence of these two male types in Scene 35. We have already discussed why women sometimes respond to intercourse by ovulating on cue (Scene 15) and why men respond to the sight of another couple having intercourse by becoming aroused and erect (Scene 9). These sexual responses were important factors in determining the outcome of the scene we have just witnessed. Perhaps most important of all were the women's decisions to drop the usual coyness and secrecy that one, if not the other, would usually have retained, in favour of blatant and open infidelity. The unconscious motivation was the same for both of them. In their lifetime of shopping around for genes, they had each reached a stage when it was worth trying someone new. As things turned out, theirs was a fair exchange.

SCENE 20

Tasteful Display

By chance, their business trip coincided with the first hot day of summer. The man had engineered the visit and they had been looking forward to it for days. The weather was a real bonus.

The woman had made a point of telling her partner where she was going and who she was going with. The last thing she wanted was to make him suspicious. He was steady and reliable, both as a partner and as father to their two children, and she didn't want to lose him. But, at the same time, she felt that she needed some excitement in her life, and the man with her in the car could be it. He was a challenge, and he was attractive. She enjoyed teasing and flirting with him – she even enjoyed arguing with him, which she did often.

In contrast, the man had not bothered to tell his partner. It was not that unusual for him to go on such a journey or to take somebody with him. In fact, his partner no longer seemed to care what he did and appeared to welcome his days away. Secretly, he had decided she was having an affair. Not that it bothered him much – in fact, he would almost have welcomed her infidelity, taking it as a licence to pursue infidelity himself.

As they drove the two hours to their destination, he pondered their budding relationship. After a year of platonic acquaintance, they knew a lot about each other. Enough, for example, to know that there was no way they could live together. Yet, somehow, even this incompatibility was attractive. Recently, they had begun to touch each other as they spoke. On a few occasions, like today, they had even greeted each other with a friendly kiss. He was sure they had an unspoken understanding. One day they would have a proper affair, snatching and engineering sex when they could. And, as he drove, he had high hopes that today might mark the beginning of that affair.

Their business completed by mid-day, they decided to enjoy their freedom and the weather and to buy food and drink for a picnic. As they drove into the country, passing field after field and wood after wood, she began to tease him. He was looking for somewhere secluded. She pretended not to realise and kept pointing out road-side verges or open fields near houses – suitable for a picnic, but nothing else. Eventually, growing hungry, she decided she had teased him long enough. When he pointed out a small, secluded field, she accepted. Rows of freshly mown hay lay silver-green in the intense sunlight. He thought it looked perfect. She said it would do.

There was a large blanket in the car which they took and

stretched out in the corner of the field, part in dappled shade, at her request, and part in sun, at his. She was wearing a loose cotton dress and, knowing from the minute she got up in the morning she could engineer a picnic, had brought a straw hat with her. He had started the day wearing a suit and tie but within minutes of finishing their business meeting he had taken off his jacket and unbuttoned his collar. Even so, sitting there in the field in the heat he looked uncomfortable. Knowing he was no stranger to nude sunbathing, she began to tease him again, challenging herself to persuade him to take off his clothes. The dénouement came when she accused him of having something to hide.

Interpreting her teasing as a sexual overture, he was becoming aroused. Even so, he felt slightly silly as he removed his shoes and socks and even sillier when he stood up, took off his shirt, then in one movement pushed down his trousers and pants. Now naked, he stood there, penis half erect, wondering what he should do next. Avoiding any reaction to his nakedness, she simply commented that he would be much more comfortable now and busied herself with the picnic. Still standing, he suggested that she, too, should take off her clothes. She shook her head, saying that she would burn, even in dappled shade.

It didn't show in her manner, but she was also becoming aroused. Wetness was beginning to ooze out of her vagina. She wanted sex with him, but was worried about losing her partner and her security. There were many things about this man that she found attractive: he was going to be one of life's achievers; she liked the way he stood up to her and sometimes even outmanoeuvred her. Not many men could do that, including her partner – these days he had become something of a doormat. But there were parts of this man's character that bothered her.

As they talked and ate, she glanced occasionally at his naked brown body and at his now shrunken penis. Sometimes, she wondered about his sex drive. The only way they were ever going to have sex would be if he engineered a situation, safe from discovery, in which she just couldn't say no. So far, in the year they had known each other, he hadn't even got close. His lack of insistence and persistence in sexual matters – at least, as far as she was concerned – was a surprising contrast to his drive and sharpness in all other areas of life. Maybe he had some sexual problem. True, he had produced two children, but maybe his problem was a recent development. All of his moves with her so far had been rather ill conceived. In fact, the way he had contrived their trip today was the nearest he had come to showing any real sexual enterprise.

Even so, whatever he did today, there was no chance of her saying yes. Not here, not in a field with stubble, flies and searing sun. She wouldn't take off her clothes and she wouldn't have sex with them on. How could she explain dirt and grass stains on her clothes or stubble scratches on her back to her partner? Anyway, she had other ways of amusing herself. Although they were evenly matched in most areas, there was one in which she could always outmanoeuvre him. All she had to do was give him the merest sniff of sex. Then, like a puppeteer, she could control him for a while. She could even get him to take off all his clothes in the middle of a field.

On the pretext of wanting to watch the movements of his naked body, she urged him to walk slowly across the field to a tree on the far side, then run back. After making her promise not to run off with his clothes, he complied. He felt self-conscious and silly. But he also felt excited. As he walked away, she studied him closely. Nice bum, she thought. One day, one day . . . but not today.

Even if she hadn't beckoned to him to return, he would have run back anyway. No sooner had he reached the far side of the field than they heard the sound of a car coming along the narrow road. Momentarily he froze. He just knew it would be a police car. He could see his career and his relationship disappearing in front of him. Caught naked in a field with another woman. In sudden panic he ran towards her. By the time he reached her side and sank panting to his knees, she was laughing hysterically. The car drove straight past.

In the stress and hilarity of the moment they each put a hand on the other's thigh as they knelt and composed themselves. This is it, he thought, and when his breathing returned to normal, leaned forward to kiss her. For a few seconds she responded, but as soon as she felt his urgency mounting she gently pushed him away. She told him she didn't want to, not here, not now. Disappointment clouded his face. She felt sorry for him, and a bit guilty.

Placing her hand on his thigh again, she promised him there would be better moments and places. One day, they would have sex. In fact, one day, she would have his baby. She watched his face, looking for a reaction. She loved melodrama.

Completely at a loss as to what to say or do he pulled her to him, hugged her long and gently, then kissed her again. This time, she thought she had better respond. As they kissed, she felt his naked penis rise against her belly. His kisses became harder and deeper. When he began to pull up her dress, she knew she had to act quickly, but as she pulled away, he resisted, holding her even more tightly. For a moment, she thought he was going to rape her.

As he let her go, she sank back on her haunches. With him still on his knees, her face was level with his now very erect penis. She took it in her hand, kissed its tip, and spoke to it softly, stroking it as if it were a cat. After a few moments, she took it into her mouth and slowly worked backwards and forwards. He tasted clean and waxy. After twenty seconds or so, she took it out, formed a tunnel with her hand, and began very slowly to pump. Then she let go, saying she wanted to watch him finish himself off.

Convinced by now that they weren't going to have intercourse, he needed little encouragement. If he didn't ejaculate soon, he would go crazy. As she watched him pump, her gaze switched from his penis to his face and back again. She watched intently, so as not to miss the moment. He came quickly, turning sideways so as not to spray her. The first two spurts shot out so fast she couldn't see where they went. Then she saw the final three spurts emerge more slowly and drip on to the grass in front of them. After he had worked out the last drop with his hand, he too sank back on his haunches.

Almost immediately it was over, he felt depressed and a little cheated. He had wanted to come inside her, not in mid-air. For her part, she felt relieved that his urgency was now gone. She also felt a genuine tenderness for his body and, she had to admit it, some pleasure at the way she had made him dance when she pulled his strings.

~

Humans are not the only mammals of which the female smells, licks or sucks the male's penis as part of courtship. Monkeys and apes do so often. So do rats, dogs and many others. Usually, such intimate activity follows through to intercourse, but not always. Sometimes (as in this scene – which involves the same couple as in Scenes 10 and 26) the male responds by ejaculating in mid-air, rather than in the female. Is this a mistake – some error in male or female programming? Or is it possible that when a male ejaculates without intercourse in the presence of a female they both increase their reproductive success?

Of course, most people would probably argue that such behaviour is simply the result of the generation of sexual excitement and its gratification. As we have already discussed (Scene 10), however, any behaviour that has evolved to be sex-ually exciting usually has repercussions in terms of reproductive success. This is as true for a man ejaculating in a woman's presence as it is for any other sexually exciting piece of behaviour.

In effect, a man seeks or allows oral sex as a display of health or fidelity. He ejaculates openly as a display of health and potency. From time to time, the benefits of such display outweigh any cost. The man in the scene, for example, gained from ejaculating in front of his potential lover. Moreover, he gained more than he would have done from saving those sperm for inseminating his partner later in the day (Scene 10).

The sperm he ejaculated were shed, not wasted: the consequences for his next ejaculate were the same as if he had simply masturbated (Scene 12). Whether a man stimulates himself or whether he is stimulated by a woman, the number of sperm he sheds is the same. Only when he has intercourse are the numbers different. Often, of course, as in this scene, a man does not know until the last minute whether he is going to have intercourse or ejaculate in mid-air. That the two ejaculates are different shows how quickly a man can adjust his ejaculate (Scene 4).

After shedding sperm, the man in Scene 20 had an even younger and more killer-rich ejaculate waiting in his tubes than he had before. If the woman did now change her mind and allow him to inseminate her later anyway, his new sperm army would be as good as, if not better than, the one he would have introduced earlier. The problem comes when he returns home to his partner. His waiting ejaculate is now right for inseminating a mistress (Scene 13), not for routine sex with a partner (Scene 12) – or at least, not a faithful partner. The point is, however, that if his partner really is faithful, he can afford to introduce into her body the occasional sub-optimal ejaculate. If she is not, he cannot.

Men's bodies seem to work to the following rule: 'If I have had the opportunity to be unfaithful, then while I was away so too has my partner.' In the case of the man in the scene, this possibility was in fact a reality – his partner had been unfaithful in his absence (Scene 10). So, when he inseminated her later that day, he had the inseminate he needed. It was the young, aggressive and fertile inseminate appropriate to immediate sperm warfare, not the defensive, low-fertility sort appropriate to routine sex. Once a man enters a phase of infidelity, therefore, similar ejaculates become appropriate for both his partner and his lover.

Having sex with two women does not prevent him from producing an ejaculate suitable for both – no matter how often he has sex with either.

So the man lost little, if anything, by ejaculating on to the grass – except, of course, that ejaculating on to grass does not fertilise eggs. He might have missed the chance of making his companion pregnant. But even that may not be as lost an opportunity as it might at first seem. If the woman didn't want him to inseminate her, there was a good chance that she was not in her fertile phase anyway. Since, as we saw in Scene 6, women are most likely to have sex with a lover during their fertile phase, a man has little to lose reproductively by leaving his mistress to decide when they should have sex.

The man's strategy was to be patient now in the hope of future rewards. In order for this strategy to pay off, his behaviour in the field today needed to increase his chances of gaining those future rewards. He didn't know it, but as the pair of them prepared for their picnic, his prospects were on a knife's edge.

The woman's body did not want his partnership, only his genes. In this new man her body saw qualities that would suit her next child, if she were to have one. She would commit herself to having his child only if these qualities outweighed the potential costs of infidelity. The equation was finely balanced, not least because her potential lover had so far given no evidence of sexual potency. From her body's point of view, this afternoon was to be a test of his potential – the final collection of facts, the final balancing of the equation. Subconsciously, she was not after sperm, but information.

First, there was his body. She had never seen him naked but, when she did, she found his body as good as she had expected. Like most men, of course, he assumed her main interest would be in the size of his penis and muscles. What in fact interested her most in his nakedness were his buttocks. The best indication of a man's health and hormones is the ratio between his waist and buttocks. Ideally, a tape round his waist should measure nearly the same (about 90 per cent or so) as a tape around his hips and buttocks. Firm, tight buttocks are a good, though of course not perfect, indication of his health and fertility.

Next, there was his ability to gain an erection and ejaculate. Not until he began to kiss her after the incident with the car was her body finally reassured that he was not impotent.

Finally, there was his sexual health. The best way to check for infection was to examine his penis closely. She looked closely, then licked and tasted him. Absence of rashes and sores and a reasonably pleasant taste are a good sign of health, and her body knew this. The ejaculate also gives away a great deal. A liquid, whitish ejaculate with a normal smell is a sign of health, whereas discoloration, particularly bright yellow or orange, or a bad smell, are often signs of infection. So, too, are traces of blood.

In those few minutes the woman collected an abundance of information about the man, much of which would have been hidden from her if she had simply had sex with him. As it happened, he passed all of the tests.

It is not only during her first encounter with a potential lover that a woman might gain from tasting a man's penis. From time to time, she may also gain from doing so during routine sex. Partners who were once healthy may become diseased. A woman benefits from being alerted to this change by seeing, smelling or tasting an unsavoury penis. She can also use oral sex to taste or smell infidelity (as can a man – Scene 10). Traces of a lover can remain on a man's penis for hours. Not only a man's partner but also his mistress can gain from oral sex. The latter will expect to find traces of her lover's partner on his penis; if she can also taste her vagina, she can either be reassured that the other woman is healthy or be alerted that she is not.

There are a number of other ways in which a woman might gain from making a man display his ejaculate – again, not only on her first encounter with a potential lover, but also from time to time as part of routine sex. As well as seeing and smelling his ejaculate, she can be alerted to possible disease by its taste. She should also scrutinise more closely a partner who unexpectedly cannot ejaculate or who produces a small ejaculate. Of course, if it is several days since they last had routine intercourse, the explanation may simply be that he has masturbated recently. Even discovering this is useful to her (Scene 13), but, much more importantly, the explanation may be that he has recently been unfaithful. The time scale for scrutiny is rather short, but not so short that a woman cannot sometimes catch her partner out. For example, many men would have difficulty ejaculating in front of their partner in the first hour after inseminating a mistress. Moreover, the volume of their ejaculate will not return to normal for twelve hours after infidelity.

Despite this scrutiny, a man can still gain from deliberately ejaculating in front of his partner from time to time. Choosing each occasion carefully so as to display a good ejaculate can be a powerful way to advertise his continuing good health. He can also reassure, or mislead, his partner over whether he has recently been unfaithful. It is testimony to the success of this male strategy that most women probably would not notice a smaller ejaculate than usual. This is precisely because most men do not display their ejaculate very often – and when they do, they pick their moment. So most women never get a chance to learn the subtleties of how their partner's ejaculate normally varies. Consequently, they never get a chance to learn how to recognise a departure from normality. Any man who does display his ejaculate too often may find it more difficult to hide ill-health or infidelity if he ever needs to. In particular, he should avoid placing his penis in his partner's mouth or displaying his ejaculate too soon after being unfaithful. It follows that a woman's best strategy is occasionally to pick her own moment to give oral sex, or to stimulate open ejaculation. The more unexpected she can make those moments, the more information she is likely to collect.

SCENE 21

An Abandoned Selection?

The girl was wearing only a bikini and a thin see-through wrap-around, but she was still hot. She fanned her face with her hand. The young man next to her in the back seat of the car saw her action. Nodding and smiling, he mimicked her hand movement in unspoken agreement. She smiled back, wishing he could speak her language better.

Under a clear blue sky, the road ahead of them snaked up and round the side of a hill. The sea was to their left, mountains to their right. She shouted her intentions to the driver and his male companion in front, then stood up, poking her head and shoulders through the sun roof. Despite its heat, the air cooled her as it rushed past. A short distance ahead, just disappearing around the first corner, was a similar car containing the other three members of their group. The only other female, a woman a few years older than herself, was also standing. The girl tried to attract her attention, but the sound of the wind drowned her shouts.

Soon, both cars pulled up in a lay-by under overhanging trees. As the noise of the engines died, it was replaced by the cicadas' loud singing. The seven people gathered together their bags, para-sols, beach mats and towels, then set off along a tree-lined path, easy and relaxed in each other's company. Six of them had been travelling together for weeks, seeing as much of the continent as they could, as cheaply as they could. Since arriving at the coast a week earlier, their days had been spent on secluded beaches. Their nights had been spent at beach bars, intoxicated by cheap alcohol, cheap drugs and good company. It was on such an evening, two nights ago, that they had met the young stranger who was now with them.

After a few minutes' walk, they arrived at the top of a high cliff overlooking sand and sea. Pausing, they took in the view, each casually putting an arm round the person next to them. The girl had one arm round the woman's waist, her other round the stranger's. In sudden exhilaration, she kissed first the woman then the man, then ran forward urging the others to follow. Once over the brow of the cliff, they had to make their way down a long, zigzag path, steep and obscure enough to deter the casual holidaymaker. They were lucky. They had inside information – the young stranger had been born in the area. Although he had travelled the world, every summer he came back to his beloved coast.

The path was very narrow. The girl, in front, did not notice the four young men walking up the path until she turned the corner and nearly bumped into them. They exchanged greetings, but she was concentrating so hard on not falling off the cliff that she failed to notice the close interest they took in her and the rest of her group. They were to change her life in a way she could never have guessed – but even had she known, she might not have cared. Elated by her surroundings and excited at the afternoon's prospects, she was feeling particularly frisky.

First to arrive on the beach, the girl took off her sandals and ran across the sand to the water's edge, soon to be joined by the others. After allowing the water to lap around their feet for a few moments the group moved on, following their local friend. They walked purposefully along the deserted beach towards tall rocks, jutting out into the sea. As they drew nearer, they could see that the rocks had 'nudista solo' crudely aerosolled in white paint, high up on their surface.

The six friends and their guide clambered over the first outcrop into a small sandy cove. Ensconced under a parasol, stretched out in the shade were two naked women. Nearby, their naked children were playing with a beach ball. A short distance out to sea a small boat-load of tourists travelled slowly past. At the top of the beach, almost hidden in the entrance to a small cave, was an old man. Wearing nothing except a straw hat, he was perched on a folding chair. One hand held a walking stick, the other rested on his genitals. He was watching the naked women closely.

The group clambered over more rocks and crossed two more coves, both deserted, before entering a third, more secluded, one. An overhanging cliff face shielded the beach from above, and a curve in the seaward rocks hid it from boat-bound voyeurs. The group threw down their towels and bags and immediately removed their scanty clothes. The local man was no stranger to nudity and had welcomed this chance to take his new friends to the nudist beach. They seemed to share his relaxed attitude to nakedness and were obviously quite used to each other's bodies. What he did not know was that at one time or another since coming on holiday, both women had had sex with each of the men, as well as with each other.

Almost within seconds of removing their clothes they were all cooling off in the sea. The younger girl began some horseplay with two of the men, splashing them, jumping on their backs, and grabbing at their genitals. They responded by pushing her under the water, then swimming strongly out to sea. She began to follow, but soon turned and swam back to the beach.

One by one, they came out of the water. Soon, they were all stretched out naked in the sun, passing around a bottle of wine and the first of that afternoon's joints. There was little real conversation, just a relaxed, drug-induced sense of well-being. The girl was lying with the other woman, resting her head on the other's stomach. As they talked, drank and smoked, they occasionally touched and stroked those bits of each other's bodies that they could reach without effort.

The women stayed like this for nearly an hour, sometimes

talking, sometimes drifting in and out of sleep. After a while, the girl began to feel restless. As she looked around, she could see that some of the men were also asleep. Sitting up and stretching, she nudged the other woman and pointed to the young stranger next to them. Fast asleep, his arms and legs were gently twitching. So too was his very erect penis. The girl whispered in her friend's ear. They smiled in anticipation, then went across on hands and knees to the man's side.

There, the girl watched and waited while the woman knelt by his head and bent forward to kiss him on the lips. He jumped, then jumped again when the girl kissed his genitals. As the two women kissed and stroked his body, every part of him relaxed – except one. And that grew even harder. After a few minutes, the girl sat astride his groin, massaging his body with her hands and gently massaging his penis with her vaginal lips.

By this time, the other four men were awake. At first, they feigned an air of casual disinterest. Watching the women's antics only intermittently, they resumed their smoking and drinking, passing an occasional good-humoured comment on the stranger's progress. Eventually, however, their erections betrayed their true feelings.

The girl, still sitting astride her prey, finally initiated intercourse with him. No stranger to this position, she slowly but expertly rose and fell as if impaled on his erect penis. As she did so, her friend bent further forward to lick his nipples, resting one of her breasts on his face in the process. Her movement lifted her bottom in the air as if in display to the man sitting behind her. Throwing down his cigarette, he proclaimed he could stand it no longer. He moved quickly across to the woman and, with the minimum of foreplay, entered her from behind. For a while, the remaining three men watched, then moved closer to await their turn.

Because of their positions, the two women were facing each other. If they had bent forward, they could have kissed. But they didn't. Instead, as intercourse proceeded, they looked into each other's eyes, sharing each other's mounting sensations. The girl was disappointed when, just as she was building to a climax, the stranger ejaculated inside her too soon and began to fade.

Having been left high on a plateau of sexual excitement, she moved off him and in one continuous movement sat astride the man lying next to her. Hardly had he penetrated her than she climaxed. He carried on thrusting, but once she stopped co-operating after her orgasm his movements became awkward. Thrusting as best he could, he was finding it difficult to build up to ejaculation. He was not helped by the man next to them who, waiting for his turn, complained he was taking too long. He was helped even less when the man put his hands round the girl's waist and tickled her ribs, making her laugh. Concentrating hard, his thrusting eventually brought him to the brink of ejaculation. Then, at the very last moment, he felt the girl being pushed to one side and his penis slipping out of her vagina. Still laughing, she fell on to her back on the sand and the other man collapsed on top of her.

The prostrate man was more than a little irritated. After all his hard work, all he had managed was to ejaculate on to his stomach. His annoyance grew when, with the girl still laughing, the man who had spoiled his fun entered her and began a long but gentle intercourse. He did his best to interfere in return, but the missionary position adopted by his usurper allowed him to resist being pushed off, especially once the girl wrapped her legs around him.

While the girl's third lover in ten minutes was slowly building to his climax, the events around her became a blur, even after she had stopped laughing. She dimly registered the previous man's attempts to interfere – and the fact that the other woman was still on all fours having sex with the same man. The one man who had not yet had intercourse seemed to be urging him to get off. He was refusing, saying he hadn't finished. Briefly, voices became raised. When her most recent lover had eventually inseminated her and removed himself, the last, frustrated, man left the other couple and moved over to her. She protested a little as he lay down on top of her and opened her legs, but he took no notice. Once inside her he was vigorous to the point of aggressive. Resigned to her situation, the girl at first didn't mind his thrusting, for she was very wet with the previous men's semen. But intercourse seemed to last an eternity and, bit by bit, she felt her vagina becoming drier. It was a relief when he ejaculated and removed himself.

Orgy temporarily over, the group relaxed and returned to their drinking and smoking. One after another, they went in and out of the sea, alternating short spells of swimming with longer spells of sunbathing, drinking and smoking. At one point, the older woman invited the local man to have sex with her, which he did. But the group's orgiastic activity did not really resume for another hour – when the girl announced she wanted the man who had not yet had sex with her. This triggered the rest to claim the same favour from the other woman.

This time round, however, the frenetic atmosphere had gone. Over the next three hours or so, all five men entered both women and all eventually inseminated them. But no intercourse was rushed; thrusting, when it occurred, was long and slow; on occasion, a man and a woman would stay just coupled together, without thrusting, while they talked, smoked or drank. Often, the men removed themselves without ejaculating, only to resume some time later with whichever female next became free and willing. There was almost continuous sexual activity, but the urgency had evaporated. Gradually, the intervals between coupling grew longer as, one by one, the men ejaculated their last and lost interest in further penetration.

The last to inseminate the girl was the local man. Thirty minutes previously, one of the others had faded inside her without ejaculating. She had gone for a swim, then returned to the beach to enjoy the last of the day's sun. It was still hot. Despite all of the afternoon's activity, her vagina was still hungry. Without a penis, it felt empty. She walked over to the prostrate local and sat astride his stomach, facing away from him so that she could talk to her female friend. Drink and joints were still being passed around, but less enthusiastically than before.

As she and the woman conversed, the girl's hands idly played with the young man's penis. When eventually it hardened she impaled herself. It was a sense of completeness, not excitement, that she felt when he was inside her. They stayed like that for about ten minutes – she conversing with the woman, he lying on his back and smoking. All the time, despite their other activities, their bodies were making subliminal movements, producing a tiny, gentle, but continuous thrusting of his penis inside her. This sensation was just what she wanted. Then, suddenly and unexpectedly, her feelings changed – she noticed the tingle of a potential orgasm. She responded first by rocking on his penis more vigorously, then by reaching down to massage her clitoris. He, in his turn, gradually increased his thrusting. At first, as sensations mounted, she closed her eyes. Then, for the second time that afternoon she stared into the eyes of the woman in front of her. The woman urged her on, then at just the right moment bent forward and began kissing her thigh. The girl climaxed, and within seconds so did the man.

An hour later, when the sun disappeared behind the cliff, the group began the long haul back up. They were all tired, both from their sexual exertions and from their various drugs. But adrenaline fired them into activity when they arrived back at the lay-by to find that their cars had gone. What they didn't know was that they had been stolen by the lads who had watched them arrive at the beach, hours earlier.

The loss of their transport in such a backwater was devastating. The cars had been their travelling homes and had contained all their luggage, as well as their passports and money. It took an hour to walk to the nearest village, and then another hour of phone calls by their local friend to alert the police and arrange a lift back to the next town.

Without their passports they had immediate trouble with the local bureaucracy, not least when they tried to obtain money. In the end they had to borrow some cash from their local friend, just enough to spend the next two days hitch-hiking to the nearest city where they could get new passports. They snatched sleep when and where they could. It was while they were all cramped in the back of a lorry that the woman first noticed how unusually withdrawn the girl had become. When pressed, she explained that her contraceptive pills had been in the luggage but that everybody had been so preoccupied with passports that she hadn't liked to mention it. The woman had no such problem herself: as they all knew, she had been having unprotected sex with many different men for ten years now and was totally convinced of her infertility. But she immediately worried for her friend. They planned to do something about it in the city, but the time taken to get their passports and deal with the bureaucracy, the language problems and their tiredness made them decide to wait until they were back on the coast.

As soon as they returned, the girl recruited the help of the local man, the first and last to inseminate her on the beach four days earlier. It was a further day before she managed to get hold of a new packet of pills. By then, the two cars but not the luggage had been recovered by the police – and the girl was pregnant. The sperm war between five different armies that had taken place inside her had produced a winner.

~

In Scenes 18 to 20 we discussed the problems women face in selecting a mate. But in the scene we have just witnessed the two women seem to give up all attempt at mate choice in favour of wanton promiscuity. What circumstances could favour such abandoned behaviour from the point of view of a woman's reproductive success, and what are the repercussions likely to be?

In most societies, relatively few people take part in an orgy in their lifetime, perhaps no more than 1 per cent according to a recent British survey of nearly four thousand women. Occasionally, however, such behaviour is more common. The classic examples, of course, are historical – for instance, the well documented orgies of ancient Rome. Anthropologically, there are some societies that have ritualised orgies, particularly for adolescents, in addition to those that may occur spontaneously. Taken as a whole, extreme orgiastic behaviour is relatively uncommon, but in a less extreme form equivalent behaviour is not so unusual.

At one extreme is an orgy in which a woman, as in Scene 21, allows several men to inseminate her not only within a short length of time but also in each other's presence. At the other extreme is conventional infidelity in which a woman allows two men to inseminate her over a slightly longer time span and not in each other's presence. At both extremes, what the woman is doing is essentially the same. Having selected two or more men (by the criteria we discussed in Scene 18) as suitable genetic fathers for her next child, she calls their sperm to battle. This ensures that her child will inherit not only all the other qualities she has selected, but also the genes for the production of a competitive ejaculate. As long as this latter benefit outweighs any associated costs, such as a greater risk of disease from having sex with two men instead of one, she will gain from her behaviour.

At first sight it might seem that such a strategy only works if the woman conceives a son. After all, a daughter will not produce an ejaculate, competitive or not. And to some extent this is correct – the sex of the child does make some difference to the woman's gain. But not much, because both sons and daughters will inherit genes, to be passed on to their grandsons, great-grandsons and so on, for all the qualities the woman has selected in the man who wins the war – including his genes for a competitive ejaculate. The only bonus from having a son is that the woman will have produced a male descendant with a competitive ejaculate in the *very next generation*.

In pursuing this strategy, a woman encounters a major problem – the men she selects to compete will rarely be given an equal chance to demonstrate their prowess at sperm warfare, because when the time interval between inseminations is relatively long, the outcome of that warfare is determined more by when she ovulates than by the competitiveness of the two armies. In Scene 6 we saw that if the woman's partner had inseminated her just a few hours earlier, he would have won rather than lost the war. Not because his army would have been any more or less competitive, but simply because his sudden flood of new sperm would have reached the egg just in time, rather than just too late.

The closer together in time a woman can procure ejaculates from different men, the better she tests their ejaculates' competitiveness. The ultimate would be if two inseminates could be mixed together in the seminal pool before any sperm were allowed to leave. Then the ejaculates would have exactly equal opportunities, and victory would go to the more competitive. But this could only happen if a woman had two men's penises in her vagina at the same time and both ejaculated simultaneously. Maybe this does sometimes happen – but not often enough to be discussed seriously.

Interestingly, this problem of timing can be turned to a woman's advantage. By adjusting the interval between inseminations by different men, she can bias the competition to favour *either* ejaculate competitiveness *or* the men's other qualities. Essentially, the closer together she allows different men to inseminate her, the more weight she gives to the competitiveness of their ejaculates; the further apart, the more weight she gives to their other qualities (especially if her body times ovulation to favour the man whose qualities she prefers). Always, no matter how much she biases the competition in favour of one male, another male can still win if his ejaculate is sufficiently competitive.

Most often, when a woman pursues a sperm warfare strategy there are hours or even days between inseminations by different men. A British survey in the late 1980s showed that, in a lifetime (by the time they have had three thousand inseminations), about 80 per cent of women have had sex with two different men within five days; 69 per cent within one day; 13 per cent within an hour; and 1 per cent within thirty minutes. This suggests that, most often, women have clear preferences based on the qualities discussed in Scene 18. Nevertheless, the survey also suggests that on occasion women *do* give priority to ejaculate competitiveness.

In Scene 21 the two women allowed, even encouraged, two or three different men to inseminate them within minutes. They also allowed five different men to inseminate them within a few hours. These women had selected five potential fathers for their children. Now, by having sex with all five in such a short space of time, they were giving as much priority as they could to the competitiveness of the men's ejaculates.

At first glance it might seem strange, from the viewpoint of reproductive success, that the woman in the scene who suspected she was infertile nevertheless pursued a sperm warfare strategy just as enthusiastically as the younger girl. We might have expected that she had little to gain from putting ejaculates in competition. We might also have expected that, knowing her suspicions, the men would be less interested in inseminating her than the girl. There was some indication of the latter, but not much. And what difference there was could have been due simply to the girl being the younger of the two (Scene 18). But the fact remains that on the whole everybody behaved as if the woman *was* fertile, including the woman herself. Why?

The explanation is that nobody really knew whether the

woman was infertile or not. She may have been, of course – one of the costs of having many sexual partners is a greater risk of contracting a sexually transmitted disease, and one of the costs of a sexually transmitted disease can be infertility (over 50 per cent of infertility cases are a result of such disease). Even so, despite her suspicions she may still have been fertile. Natural contraception (Scene 16) can make a woman appear to be infertile, maybe for years. But then, with the right man in the right circumstances, she may conceive.

Whether the woman in the scene was infertile or not, her behaviour will have been unaffected. All her body can do in its pursuit of reproductive success is assume that it really is just a question of being inseminated by the right man at the right time, and motivate her accordingly. Failure to conceive may even make a woman *more likely* to behave like the woman in the scene. Not many seemingly infertile women take this promiscuous route to conception, but some do. The woman's behaviour was just as much a reflection of her body's pursuit of reproductive success as was that of the fertile younger girl. So too was their bisexuality (Scene 31).

By taking part in an orgy, the two women were testing the men as much as possible for differences in prowess at sperm warfare. Although in large part this prowess will depend on the competitiveness of their ejaculates, a number of other factors will inevitably have influenced it – some more than others.

At first sight, for example, some intercourse positions might seem to be better for sperm retention than others – in which case, males who manage to engineer the best position will have most sperm retained and thus most chance of winning the war. As it happens, coupling position probably makes little difference to the eventual outcome. The seminal pool is deposited at the top of the vagina no matter which position is adopted. As we saw earlier, the pool stays there as the penis is withdrawn, partly because it quickly coagulates around the cervix and partly because the vagina closes behind the withdrawing penis, holding the pool in place. Only in the woman-on-top position is there a danger that part of the pool may be lost before sperm have had time to escape into the cervix. Even then, this is only a danger if the man withdraws too quickly after insemination.

Not only does coupling position have little influence on the retention of the seminal pool, it also has little if any influence on the ability of sperm to leave the pool and proceed to the cervix. This is because of the cervix's neat design. In the missionary position, for example, the pool is deposited on the floor of the vagina with the cervix dipping into it (Scene 3). In the rear entry position, the cervix is either underneath the seminal pool, like a plughole in a sink, or it sticks up and hangs back down, like a coiled spring 'walking' downstairs. In the woman-on-top position, the cervix sticks out sideways and hangs down into the pool. Moreover, no matter how a woman changes her position after insemination, gravity will ensure that even a coagulated pool of semen will slide into a new position. Gravity will also ensure that her cervix continues to dangle into the pool, maintaining contact between mucus and semen.

The main differences between coupling positions concern not their influence on sperm retention but their influence on a couple's vigilance and vulnerability during intercourse. As one of the men in Scene 21 discovered, it is much easier to defend some positions than others. Moreover, the rear-entry position allows at least the man to be more vigilant (Scene 34).

As far as sperm warfare is concerned, much more important than the position a man adopts for intercourse is the size of the sperm army he introduces into the woman. All of the men should have introduced as many sperm as possible, as it should have been clear that warfare was imminent. Failure to inject large numbers would have guaranteed a very poor performance.

However, the men did have a problem of strategy because there were two women to inseminate. Whereas they should have put a larger army into their first woman because a chance to inseminate their second woman could not be guaranteed, they should have avoided putting all of their sperm into the first because an opportunity to inseminate the second did seem likely. Precisely how many sperm they deployed will have depended in part on how long it was since they last had sex with each woman. Suppose it was forty-eight hours. In such a circumstance, we should expect the average man to inject about 450 million sperm into the first woman and about 350 million into the second. Not all men, however, are average.

If any of them had been a sperm war specialist, such as we met in Scene 19, he would have had an immediate advantage. His large testes and huge sperm armies could have won the day. But there could have been another anatomical feature separating the men – penis size, as well as testis size, could have made a difference.

Most people see the human penis as functional rather than aesthetic. What most people don't realise is that there is more to its function than simply delivering sperm to the top of the vagina. The human penis is a very effective suction piston. Its shape is no accident, nor is the backward and forward thrusting that accompanies penetration. The penis has evolved its size and shape to remove any material that is already present in the woman's vagina. In humans, it is particularly effective at removing any seminal pool or any as yet unejected flowback that may still be present. As the penis pushes forward, with any foreskin now pulled back off the glans on to the shaft, the smooth, blunt tip of the glans pushes through any semen or mucus in the vagina. As it then begins to pull back, two things happen: any material behind the vertical flanges at the back of the glans gets dragged back down the vagina; and any material ahead of the penis gets sucked further down the vagina, ready to be pushed through at the next forward thrust. Rapid backward and forward thrusting during intercourse thus sucks out any seminal material from a recent insemination. It may even help to remove some mucus and blocker sperm out of the cervix. The longer and more rapid the thrusting, the more the vagina is sucked clean of any previous insemination. The larger the penis, the more effective the removal.

As each man in Scene 21 tried to claim the prizes in front of him, success or failure depended in part on how he adjusted his behaviour according to his position in the queue for each woman's body.

The stranger, the first to inseminate the younger girl, was in pole position – the man with maximum opportunity to deploy his army and make life difficult for all who followed him. Even if he had a chance to inseminate the other woman later, his situation would be less favourable. More than any of the other men, he had most to gain from injecting more of his sperm now (say six hundred million), and saving fewer (say two hundred million) for later. Because he was in pole position, he also needed to ejaculate quickly. Any delay might have given another male the chance to physically oust him and deposit his own sperm inside the girl first. It was because of this urgency that he ejaculated before the girl climaxed.

Having succeeded in being first to inseminate the girl, what the stranger should then have done, but didn't, was try to delay the next man as long as possible. That way, he would have given his army maximum time to leave the seminal pool and deploy itself. (The man who was first to inseminate the older woman succeeded in doing just that, despite triggering aggression.) The girl the stranger had inseminated, having failed to climax while he was inside her, had other plans. She moved immediately from the stranger to the next man.

This man needed to get his penis inside the girl as quickly as possible after the stranger had finished – the sooner he could do this, the sooner he could pursue his own best strategy. He would, of course, have been helped in his urgency by having watched the girl's previous intercourse. Like most men, he would have found the sight of a copulating couple sexually arousing (Scene 9). While he was waiting his turn his penis would have been erect and ready for action. Once inside her, in fact, he had a choice of strategy. He could have ejaculated really quickly, reducing as much as possible the stranger's head start, but the disadvantage of this option was that he would have shot his sperm straight into the stranger's seminal pool. So instead, he tried thrusting long and vigorously to remove that seminal pool first – and paid the price. Because of his delay, he was displaced and relegated from the possibility of being second to not inseminating her at all – at least, not in the first round.

Just as it is better to be first to inseminate a woman than second, so it is better to be second than third. Rather than wait his turn, therefore, the third man successfully engineered events to inseminate the girl second, not third. At the risk of aggression from the man he displaced, he moved in at just the right moment.

The fourth man then had no real alternative. Having been thwarted in his attempt to be the second to inseminate the other woman, he had to settle with being the third to inseminate the girl. Her vagina already contained a seminal pool from two different men. His only option was to try to remove as much of that pool as possible by long, vigorous thrusting. In this he succeeded, for the girl eventually felt her vagina becoming dry. So when he ejaculated, his pool had the girl's vagina to itself. But the cost of removing the pool was that he lost time, and sperm from the first two men had precious extra seconds to swim into her cervix and prepare an ambush for the fourth man's army.

It was because the best strategy for all the men was to try to inseminate each woman as soon as possible that the first round of sexual activity was frenetic and mildly aggressive. The second round, though, was quite different. As the afternoon wore on, both pace and tactics changed. Why?

It is important to realise that winning the first sperm battle inside the girl did not necessarily guarantee claiming the prize of fertilisation. It would do so *only if* she ovulated some time within the following two days, during which her tract would be monopolised by those first few armies. If she ovulated not two but four or five days after the orgy, the situation could be very different. So having done the best they could to win the first skirmishes, there was some value to the men in a longer-term back-up strategy.

By four or five days after the orgy, most of the sperm from the first skirmishes would be dead or dying. As time passed and ovulation still hadn't happened, the chances of fertilisation being achieved by a sperm introduced into the girl relatively late in the orgy improved considerably – especially as these later sperm would also have been very young. They would be the last to lose their fertility inside the girl, which meant that if ovulation was delayed they would have a good chance of claiming the egg. Of course, such late sperm would have to run the gauntlet of the earlier blockers and killers, but nevertheless they might just succeed.

The fact that each man divided more or less all of his initial store of sperm between the two women during his first two inseminations would not prevent him from pursuing a back-up strategy. A man adds new, young killers and egg-getters to the back of his sperm queue at the rate of about twelve million per hour. In the five hours or so from first insemination to last on that abandoned afternoon, each man will have mobilised a further sixty million young sperm. The longer each waited to make his final insemination of each woman, the more young sperm he would have available.

In the second phase of activity, the men were trying to do two things. First, they were trying to remove as much of previous seminal pools as they could. They were going to inseminate few sperm anyway and so needed as easy an escape route into the cervix as possible. Secondly, they were trying to be the last to inseminate each female with as many *young* sperm as they could manage. Their problem was that if they inseminated either woman too soon, they would have fewer sperm and might in any case not be the last inseminator. If they waited too long, and the woman lost interest, they might miss their chance altogether. Each man, once coupled to a woman, stayed in position as long as he could, thrusting as slowly and for as long as possible, his body trying to judge the best time to deposit his ejaculate. If he decided it was too soon, he would withdraw without ejaculating and wait until later. One by one, each man made his decision and ejaculated his last.

Consciously, of course, none of the men knew anything of what their bodies were trying to do. As far as they were concerned, they simply felt different levels of excitement as one by one their opportunities came to enter and inseminate one or other woman. When their bodies judged rapid insemination to be best, they experienced intense excitement and virtually spontaneous ejaculation as soon as they were inside. When their bodies wanted to remove as many traces of the previous ejaculates as possible, they experienced only just enough excitement for erection. They had to work very hard at thrusting before their bodies finally gave them the tingle of loading and the climax of ejaculation. During spells when it was best to wait a while, they temporarily lost interest and their erections dwindled. Finally, when their bodies judged there was nothing to be gained from introducing further sperm into either woman, they lost interest totally.

In this complex interplay of reactions and interactions, the man with the best chance of victory would be the one whose body best timed his own inseminations, most interfered with other men's inseminations, and best judged when and when not to couple and to thrust. It would be this man that the women's bodies were seeking to father their child. The women's role, in trying to further their own reproductive success, was to give each man maximum opportunity to prove himself – and maximum chance to make mistakes. In the first round of activity, the women benefited most by allowing the men their urgency and their thrusting. In the second round, they benefited most by allowing the men to play musical chairs with their vaginas.

In the second round, giving each man maximum opportunity to couple for long periods and to choose his moment, right or wrong, to ejaculate for the last time was a shrewd method of selection. By allowing such lengthy coupling without any other overt sexual activity, each woman was testing each man's ability to guard her vagina from other men, maintain an erection, and judge the best moment to ejaculate – all qualities she would like in sons and grandsons.

During this long, slow, relaxed second phase, the women's bodies will have experienced satisfaction simply at the feel of an erect penis inside them. Not until each body decided it had given the men maximum opportunity to show their prowess at sperm warfare would it have lost interest in simply coupling. During the short, frenetic first phase, though, and once or twice during the second, their bodies will have looked for a different satisfaction. Instead of the quiet, gentle feel of a penis in their vaginas, they will have looked for the right moment to begin a headlong rush into orgasm.

We do not know whether the older woman in the scene climaxed or not. The girl climaxed twice; once immediately she was entered by the second man and once right at the end, while she was impaled on the stranger. These two climaxes actually meant that, despite everything, she was still not giving all the men an equal chance at winning the sperm war inside her. She was still showing favouritism – and the person she favoured was the stranger. *How* she implemented this favouritism won't be clear until Scene 25. For the moment, the question of interest is *why* she preferred the stranger.

Xenophilia, a preference for strangers, is a powerful factor in a woman's choice of mate, particularly as a target for infidelity.

In this, women are typical primates. A female red monkey, for instance, was observed allowing nearly every new male she met to inseminate her, while avoiding those she knew. Similarly, female macaques are known to go out of their way to allow male newcomers to their troop to have sex with them. They do this even though such newcomers occupy a very low rank and are often victimised by the resident males.

Of course, a female primate does not allow every male stranger to inseminate her. The males still have to satisfy her other criteria for mate choice. The point is that, if two males satisfy her criteria equally, she is more likely to prefer the stranger. It is generally thought that what the female is doing here is taking precautions for the future. If the new male eventually achieves a powerful position in the troop's hierarchy, he is more likely to behave favourably towards her and her offspring if there is some chance he is the father.

We cannot tell whether the girl in Scene 21 preferred the stranger simply because he was a stranger, or because he satisfied her criteria better than the men she already knew. Whatever the reason, she did favour his sperm. First, she solicited sex from him both before and after any of the others. Secondly, she climaxed with him, and only him, during the second phase of sexual activity. It might seem contradictory, therefore, that she climaxed not with him, but with the man following him, during the first frenetic phase. But this is not in fact a contradiction – as will become clear in Scene 25.

8

The Climax of Influence

SCENE 22

Finger on the Button

The woman said goodnight and closed the door to her young daughter's bedroom. Going into her own room, she undressed, put on her dressing-gown and went downstairs to pour herself a coffee. Her partner was out, the house was quiet, and she had at least two wonderful, peaceful hours entirely to herself. She switched on the television, picked up a magazine, and began idly flicking through the pages. Bliss.

Fifteen minutes passed, the television just a quiet noise in the background. For all the attention she gave it, the magazine, too, might as well have been in the background. The occasional image or headline caught her eye as she mechanically turned the pages, but essentially she was lost in her own thoughts – like butterflies, dancing around in her head, chasing each other in and out of her mind.

She wasn't sure which came first, the thought that she could, or the tingle between her legs that told her she should. But suddenly the idea was there. It would be nice, but could she be bothered? A few more minutes went by. She put down the magazine and stared at the fire. Maybe she would. Even having decided, she waited until she had drunk the last of her now cool coffee. On her way upstairs she took the phone off the hook. Quietly, she walked past her daughter's bedroom to her own, and bolted the door.

For a moment, she toyed with the idea of doing something different this time. Maybe she would look at the pictures she had secreted away, but she couldn't quite remember where they were. Maybe she would put something in her vagina, but she couldn't think of anything quickly to hand. Two or three years ago, she had used the handle of her partner's table tennis bat, but then had spent days imagining splinters and germs. One day, she would buy herself a vibrator, if she could only think of somewhere to hide it.

In the end she decided not to bother with anything and simply to follow her usual routine. Lying down on the bed, she untied the belt of her dressing-gown so that it fell open. Her left hand went across to her right nipple. Her right hand went first to her mouth to collect saliva on her fingers, then down between her legs to her clitoris. She began to masturbate.

At first, she had difficulty focusing on her sensations. Other, random, non-sexual thoughts kept intruding. She tried several fantasies. Old favourites like being undressed, licked and caressed by another woman or being worked on by two men got her started, but didn't take her very far. Then, after about five minutes of stimulation, a fantasy scene with a friend and her partner finally started her sensations buzzing. Lubricant was spreading over her genitals and her fingers. Her breathing grew heavier and her heart began to race. The images had now gone and her focus was entirely on the sensation between her legs, under her moving fingers. As she massaged her clitoris more and more vigorously, first this way, then that, first with this rhythm, then with that, she felt the flush spreading over her chest, throat and face. Her body tensed, poised on the brink. Just one more touch of her moist, swollen clitoris, and she was there. In silent climax, her thighs and genitals spasmed, quickly at first, then at increasingly long intervals. Finally, she relaxed. It was done.

Not bad, she thought. Maybe 7 out of 10.

After lying on the bed in post-climactic torpor for a few minutes, she stood up and re-belted her dressing-gown. Downstairs, she put the phone back on the hook, poured herself a glass of milk and returned to the warmth of the sitting-room and the drone of the television. Warming herself in front of the fire, she tried to remember the last time she had masturbated. It didn't happen that often, perhaps only three or four times a month. Maybe the last time was about ten days ago. When she did masturbate, the orgasm was usually fairly good. Certainly better than any her partner gave her. Occasionally, he could give her a 5 out of 10 with his hands, but with his penis it was usually about 2 out of 10. Actually, it was usually 0 out of 10 because she rarely climaxed during intercourse.

Masturbation had first added sexual colour to her life in her late teens. At first, her orgasms were pathetic, the merest tingles. She scarcely knew when she had climaxed – 1 out of 10 at the most. By the time she was twenty, however, she was often hitting 7.

Her partner didn't know she masturbated. In fact, the only person who did know was the woman who, with her partner, had just starred in her fantasy. They had been friends for years, having met long before they had each met their respective partners. In one drunken conversation when they were twenty they had talked about masturbation. She had felt quite daring when she admitted that yes, she did 'sometimes', but was then totally deflated when her friend announced that recently she had been masturbating virtually every night. If she didn't, she claimed, it was difficult to sleep. This had seemed particularly surprising because at the time her friend had been suffering from a prolonged bout of cystitis. The disease had invaded her left kidney and was proving difficult to shift.

For a week after that conversation, she had attempted to emulate her friend and masturbated every night. However, two nights in a row seemed to be her limit. Even then, the second night was hard work. Quickly, she had reverted to the once a week or so that came naturally.

When her partner returned from his night out, she was curled up on the settee, watching television. They exchanged pleasantries and after a while began to prepare for bed. As they lay down, she felt like intercourse. The thought of a penis inside her was mildly appealing. She tried a half-hearted flirtation but her partner responded with that edge of irritability which by now she knew meant 'no chance'. Mentally shrugging, she turned away from him and within minutes was asleep.

~

Of all aspects of human sexuality, the female orgasm has probably been the most enigmatic. Women differ considerably in their experiences (Scene 36). Some never have an orgasm. Others have orgasms but never during intercourse. Yet others have an orgasm nearly every time they have intercourse. Although clearly a sexual phenomenon, the female orgasm (unlike male ejaculation) is not necessary for conception. Women who never orgasm readily conceive.

Variability is one of the major problems in discussing the female orgasm. Every orgasm is a bit like every other orgasm, a bit like most other orgasms, and a bit like no other orgasm. Moreover, no two women have precisely the same range, frequency and pattern of orgasm. Women differ far more than men over when, how and how often they climax.

The five scenes in this chapter explore the ways in which orgasm and – equally importantly – the avoidance of orgasm can enhance a woman's reproductive success. Each scene illustrates a situation which will be very familiar to some women, and totally beyond the experience of others. In each situation, the function of orgasm will be the same for all women who orgasm in that particular situation. But when and how often they make use of that function will differ from woman to woman – for very good reason. These differences between women are extremely interesting and in Scene 36 we shall discuss them in their own right. In these five scenes, though, we are concerned primarily with the way that orgasm has a different influence on a woman's reproductive success depending on the situation in which it occurs.

In some ways, it is difficult to see any function in a piece of behaviour that is so variable and unpredictable – which is why many people have concluded that there is no function to orgasm other than that of giving pleasure. But as we have already seen, pleasure is not in itself a function (Scene 10), but a by-product of function. Basically, whenever the body is intent upon a particular course of action, it generates an urge to perform that action. As that urge is gratified, the sensation generated is pleasure. The female orgasm is pleasurable *because* it has a function. A woman feels like an orgasm whenever her body judges it will *enhance* her reproductive success. When her body judges it will *reduce* her reproductive success, she feels no such urge. We begin, perhaps surprisingly, not with orgasms associated with intercourse (Scenes 24 to 26) but with orgasms generated by masturbation.

Nearly 80 per cent of women masturbate to orgasm at some stage in their lives and for most, like the woman in the scene, it is a routine but not very frequent activity. The average rate of masturbation is just under once per week – slightly higher in the week or so before ovulation, slightly lower at other times. A woman is also more likely to masturbate, and to masturbate more often, as she gets older, at least up to the age of forty.

The commonest method of female masturbation is to stimulate the clitoris, usually with just the fingers. Sometimes fingers are pushed into the vagina but this is most often a complement to clitoral stimulation, not a replacement. On occasion, a penis substitute may be inserted in the vagina – again, as a complement to clitoral massage. In different cultures, objects as diverse as reindeer muscle, fruit, vegetables and commercial vibrators may be used as masturbation aids. Women have even been known to recruit a cat or dog to lick their genitals during masturbation. Even more rarely, a woman may allow a dog or some other animal to have intercourse, again as a form of masturbation.

Because of the confusion that has surrounded the function of the female orgasm, similar confusion has surrounded the role of the clitoris. This organ develops in the female foetus from the same cells as does the penis in the male foetus. All female mammals have a clitoris, the tip (or 'bulb') of which contains many nerve endings and is very sensitive – more sensitive, in fact, than the tip of the penis.

In some monkeys, mainly those from South America, the clitoris is as large if not larger than the penis. Often, it is also grooved and carries urine away from the female's body during urination, much as the penis does in males. Such large clitorises could play an active role in intercourse. They are not stimulated directly as a male thrusts, because the rear-entry position usual for primates means that the clitoris is forward of the point where the penis enters the vagina. Nevertheless, because of its size, the clitoris is relatively easy for both the male and the female to find and handle during intercourse. This is very different from the situation in most other monkeys, as well as in apes and humans, in which the clitoris is much smaller than the penis and is often concealed in folds of skin.

Inevitably, a small clitoris is even less likely to be stimulated during rear-entry intercourse. Even species which sometimes couple frontally (some monkeys, orang-utans, chimpanzees, gorillas and of course humans) do not necessarily stimulate the clitoris during intercourse. In fact, in such species the clitoris is so cunningly positioned that it actually seems designed *not* to be automatically stimulated by intercourse. Moreover, small, semi- or totally concealed clitorises are difficult for a male to find. The overriding impression, therefore, is that a small, concealed clitoris like that of a woman is primarily a push-button for self-masturbation – not an organ for *automatic* involvement in intercourse (Scene 25).

Women are not the only female mammals to masturbate. Many other primates have been seen stimulating their clitorises, either with their hands or by rubbing their vulvas on the ground or on branches. Chimpanzees have been seen inserting leafy twigs in their vaginas, then brushing them against vertical objects, causing vibration. Nor is female masturbation only found in primates. Female porcupines have been seen to stand astride a stick then run, causing the stick to vibrate and so stimulate the clitoris.

Anecdotes apart, we know relatively little about female masturbation and orgasm in species other than humans, though we do know that a female chimp will climax if someone massages her clitoris. Moreover, once the chimp has discovered this substitute for masturbation, she will periodically solicit clitoral massage by presenting her rear-end to the person's hand. A cow also will climax if someone massages her clitoris. A few minutes after massage begins, the cow's cervix is seen to gape open and move.

The same cervical movements are seen following clitoral stimulation in humans. Film taken by a tiny camera placed inside the vagina of a masturbating woman shows that when she climaxes, her cervix also gapes, dipping into the vagina at the same time. This gaping and dipping is sometimes called 'tenting' and may happen several times during the course of a single climax. Orgasmic tenting of the cervix, combined with climactic events inside the cervix itself, are the key to the function of masturbation, for they have three main consequences.

First, masturbation (via orgasmic tenting and so on) temporarily increases the flow of mucus from the cervix into the vagina. Mucus flows slowly from the cervix all the time, but orgasm speeds it up – almost equivalent to the difference between a 'runny' nose and sneezing. When a woman becomes sexually aroused during masturbation, glands at the top of her cervix increase their rate of mucus production, particularly if she climaxes. This quickly increases the size of the mucus 'glacier' flowing through her cervix (Scene 3). At first, this is accommodated by the tenting cervix. Then, when tenting ceases, a part of the older section of the enlarged glacier is squeezed into the vagina, like toothpaste out of a tube. This flood of mucus lines her vaginal walls, laying down a thick film of lubricant ready for her next intercourse (Scene 3). This shedding of mucus, however, does more than simply enhance vaginal lubrication. Because it is the oldest section of mucus that is ejected, it carries with it much of the cervical debris, including old blocker sperm and disease organisms. This is an effective way of fighting infection. In Scene 22 the woman's friend was driven to masturbate nightly during an episode of urinogenital infection.

Secondly, masturbation increases the acidity of the cervical mucus. When a woman's cervix tents, the mucus inside stretches sideways. New channels form through the mucus and at the same time material from the top of the vagina is effectively 'sucked' into the tip of the 'elephant's trunk' of the cervix (Scene 3). The fluids at the top of the vagina are very acidic, and as the cervix tents several times during the course of the climax, acid spreads through the mucus channels making the older portion of the cervical glacier more acidic than it was before. When the climax ends, part of this acidic mucus (plus any disease organisms it contains) will be ejected back into the vagina as described above, but some will remain in the cervix. Neither sperm nor bacteria function properly in acidic mucus. For a while after masturbation, therefore, and maybe even for days, sperm are less able to swim through the mucus channels and disease organisms are less able to invade and multiply.

Thirdly, masturbation changes the strength of a woman's cervical filter – most often strengthening it. This is due not only to the increase in acidity we have just discussed but also to orgasm triggering about half of the cervical crypts (Scene 4) to void their load of sperm from the last insemination. This strengthens the filter because these sperm, on re-entering the mucus, block new mucus channels. Moreover, because many of these new blockers are high up in the cervix, their influence may last for days – persisting until the cervical glacier carries them down into the vagina. Sometimes, masturbation does *not* increase the strength of the filter. The critical factor seems to be how many of her cervical crypts still contain sperm.

The more crypts that contain sperm when a woman climaxes, the more are injected into the mucus channels and the stronger her resulting filter. Consequently, masturbation twenty-four hours after intercourse, when many crypts contain sperm, strengthens the filter more than masturbation forty-eight hours after intercourse, when fewer crypts contain sperm, and so on. As long as enough crypts still contain sperm, a second masturbation a day or so after the first strengthens the filter still further, but a third or fourth usually has no additional effect. Presumably, after the second masturbation, few if any crypts contain sperm. Similarly, if the crypts are empty of sperm for other reasons (because, for example, it is more than eight days since last intercourse or the couple used a cap or condom at the previous intercourse), masturbation again does not strengthen the cervical filter.

We have already noted the tendency for a woman to feel like masturbating more often during the fertile phase of her cycle than at other times. This now makes sense – this is precisely the time when she has most to gain from preparing her vaginal lubricant and cervical filter for intercourse. All the same, she may still feel the urge to masturbate at other times. Of course, even during infertile phases she still needs to combat disease, prepare her vagina for intercourse and adjust her cervical filter. She also still needs to confuse her partner (Scenes 23 and 26).

Masturbation is not the only way a woman can combat disease and prepare for her next insemination via orgasm. There is another way, not as common as masturbation, but experienced by a number of women. This source of orgasms is the subject of Scene 23.

SCENE 23

Dark Secret

The purple haze hanging in the air seemed quite natural, as the woman walked around the supermarket. She was alone. Ahead of her, stretching for miles into the distance, the meat aisle was completely empty of people. She went slowly past the beef, then even more slowly past the chicken, idly wondering what to cook that evening. Just beyond the bacon she stopped by a big display of grinning pigs' heads. This section was refrigerated and should be ice cold. So why was she so hot?

For a while she stood with her legs apart, trying to let cool air reach her hot, tingling groin. No sooner had she thought of taking off her knickers than it was done, and they were hanging over the front of her trolley. An attractive woman in shop assistant's uniform suddenly appeared in front of her. Picking the underwear off the trolley, the assistant held them in the air.

'Why did you take these out of their pack?' she asked.

'I didn't, they're mine. I was hot, so I took them off.'

The assistant looked at the knickers, then pressed them to her face. 'Mmm, I like the smell,' she said, her beautiful eyes peering at the woman over the flimsy garment which she still held over her nose and mouth.

'If they're yours, you won't be wearing anything under that skirt, will you?' she said. She breathed deeply through the woman's knickers a few more times, then continued, 'And you will smell just like this.' She dropped the knickers. 'Let me see if they're yours. Lift your dress and let me smell you.'

The woman refused and backed away. As if from nowhere, a crowd gathered. Standing in a semicircle with the assistant at the front, they began to hem her in. She tried to back away still further, but found herself against a wall. A man in the crowd came forward, pushed himself behind her and pinned her arms so that she couldn't move. Her heart was beginning to race and she was having difficulty breathing. She tried to speak, but the man put his hand on her throat.

The assistant began to undress, handing first her uniform then, one by one, her other garments to someone in the crowd. Her still beautiful but now demonic gaze never strayed from the woman's eyes as she gradually revealed more and more of her body. The woman startled and the crowd gasped as she took off the last of her clothes. Her top half was that of a beautiful woman, with perfect breasts. But her bottom half was that of a man – a very excited man.

The naked hermaphrodite stepped towards her. As it did so, the crowd began to chant, 'Rape her!' The woman tried to struggle but the man still held her tightly. In one movement, the hermaphrodite tore off the woman's dress and pressed itself against her – breasts against breasts, penis against groin. The beautiful female face began to kiss her passionately as the crowd's chants grew louder. The woman tried to struggle, rocking from side to side. Every so often a tiny noise escaped her throat as she tried to scream. She was going to be raped.

After kissing her hard on the lips, the face moved down and began to bite her nipples. More sounds escaped as the woman struggled still harder. Her heart felt as though it was going to jump out of her body. The face left her nipples and began to lick its way over her belly towards her groin. Then, just as it was about to give her oral sex, she climaxed. In the middle of her contractions, she woke up.

As she surfaced from her sleep, the woman briefly registered what had happened. It was a month since she had last climaxed at night. In fact, it was probably a month since she had last climaxed on her own at all. This had been a good one – 9 out of 10, she thought. She looked sideways at her snoring partner, then turned over. As her body made its descent back to normal, she drifted back into sleep.

~

We have already discussed the ways in which 'wet dreams' are important in a man's battle for reproductive success (Scene 14). Here, we consider whether nocturnal orgasms are equally important for women.

Although all women have sex dreams, not all women have nocturnal orgasms. By the age of twenty, only about 10 per cent have ever experienced a 'nocturnal'. Some may have their first as late as the age of forty. The majority of women, however, never experience (or at least remember) such an orgasm. Over a lifetime, only about 40 per cent will ever share the experience of the woman in the scene, but for those who do, nocturnal orgasms will be an important part of their sexuality. In fact, for the women who have them, nocturnals usually produce the strongest climaxes of all. Sometimes, though, the momentary waking up just at climax can prevent them from being as satisfying as a masturbatory orgasm can be.

Nocturnal orgasms are nearly always associated with dreams, although these dreams are not always sexual. Even when they are, the sexual element sometimes makes only a last-minute entrance into the dream. Often it seems that the impending orgasm triggers the sex dream, rather than the other way round. In the dream just recounted, the woman was sexually aroused even before she met the shop assistant. The climax was already on its way and the dream was manufactured by her brain simply to accompany her body's build-up. It is unusual to dream of hermaphrodites. Most often, people dream of situations that reflect their own usual sexual activity, though lesbian scenes for women who have never had a lesbian relationship are relatively common – more common than homosexual scenes for hetero-sexual men.

Women are not the only female mammals that climax spon-taneously in their sleep, though information for other species is scarce. The clearest observations are of female dogs. When bitches on heat are asleep, they occasionally become very agi-tated, presumably dreaming. On some of these occasions, mucus wets their vaginas and they have rhythmic contractions of their genitals, just like women.

The functions of nocturnal orgasms seem to be identical to those of masturbatory ones. Both types help a woman's body in its battle against infection. Both prepare her vagina for its next intercourse by depositing lubricant. And both strengthen her cer-vical filter, as long as there are sperm in her cervical crypts. In fact, no physiological differences can be detected between the two types of orgasm. Not surprisingly, therefore, nocturnals show a link with the menstrual cycle similar to that shown by mastur-batory orgasms.

In the same way that female dogs are more likely to have nocturnals when on heat, women are more likely to have noctur-nals during, or at least at the beginning of, their fertile phase. The most likely time is about a week before ovulation – precisely when all of the benefits of nocturnal and masturbatory orgasms are greatest. This peak more or less coincides with the timing of the peak urge to masturbate, but is more clear-cut for nocturnals. Women who use the contraceptive pill show no such peak in the menstrual cycle for either type of orgasm. This indicates that nocturnals and the urge to masturbate are largely under hor-monal control.

A consequence of this link between ovulation and the timing of nocturnals and masturbation is that these two types of orgasm have a seasonal peak. As we have already seen, because women are more likely to ovulate in some months of the year than others, humans are more likely to be born in some months than others (Scene 15). As nocturnals and masturbation are more likely to occur in the week or so before a woman ovulates, they also are both more likely to occur in some months than others. For example, in Britain there are peaks of birth in February/ March and September, peaks of ovulation in May/June and December, and peaks of nocturnals and masturbation in April/ May and again in November. And as in the menstrual cycle, the peaks are more pronounced for nocturnals than for masturbation.

Although there are no major physiological differences in the function of nocturnal and masturbatory orgasms, there are some minor strategic differences. Most of these derive from the relative ease with which a woman can keep the two types of orgasm a secret from her partner.

In Scene 14 we concluded that the number of wet dreams decreases as a male ages and obtains a partner, because wet dreams are *more* difficult to hide from a partner than masturbation. The opposite is probably true for women – nocturnals are *less* difficult to hide from a partner than masturbation. In Scene 23, the woman's partner was asleep when she climaxed. But even if he had been awake, he still could not have been certain that she was having an orgasm rather than simply dreaming. In contrast, he could scarcely have misinterpreted her intentions had he caught her with her hand between her legs and a banana in her vagina!

It is probably because nocturnals are more cryptic than masturbation that they are also more closely linked to the menstrual cycle. Given that both types of orgasm are advantageous if they occur at the beginning of the fertile phase, nocturnals are probably the better option – being less likely to betray the woman's fertile phase to her partner. It is probably also because nocturnals are cryptic that women, unlike men, are *more* likely to have them as they age and obtain a partner, not less.

Many women have both types of orgasm, but differ in whether they have more masturbatory orgasms than nocturnals or vice versa. Although the triggers and urges for both types are to a large extent under hormonal control, they are also influenced by external situations – one of which is potential infidelity (Scenes 6 and 17), as we shall soon discuss (Scene 26). Together, the two types of orgasm help a woman's body to give an advantage to the man it would most like to win sperm warfare. Her body's potential to influence that warfare, however, does not end with masturbation and nocturnals.

When she is next inseminated, a woman still has a variety of options. Everything now depends on whether she has an orgasm during intercourse (Scenes 24 and 25).

SCENE 24

Another Successful Failure

'Did you come?' the man asked, panting, as, propped on his arms, he held himself above her body.

'Nearly,' she replied, kindly. 'I thought I was going to, but then it just went and I couldn't get it back.'

The man slowly withdrew, then more or less collapsed by her side. 'I thought you did, just before *I* came,' he panted, his intonation a mixture of disappointment and irritation.

'No, not really. But I did enjoy it. It was nice to feel so close.'

The couple relaxed in their usual post-coital embrace and sank back into their own thoughts. He couldn't see what the problem was. This time, he had really gone for it. He had taken his time over foreplay and there was no doubt that she had become excited. Maybe she wasn't as wet as she could have been, but she was flushed. A few more seconds playing with her clitoris and she would have come, he knew she would. It had been putting in his penis that had stopped her. As he had climbed on top and pushed himself in, she had visibly lost interest – her excitement level had immediately dropped about twenty notches. Even then, he had done his best to get her back. He had thrust as long and as considerately as he could – not too hard, not too slow. He had delayed as long as he could. Really, though, he had known they weren't getting anywhere. He could sense that she was becoming bored and impatient rather than excited. In the end, he had given up and just concentrated on making himself ejaculate. It was wishful thinking rather than a lack of awareness that had made him suggest she had come just before he ejaculated. Or had he been giving her an opportunity to lie?

As far as he could see, it had to be his partner who had the problem. A year or so ago he had had a brief affair with a younger girl. She had come nearly every time. As long as he had given her a bit of foreplay and as long as he didn't ejaculate too quickly, she would come. It was as easy as that. His partner, however, rarely came while he was inside her. Once or twice a month, perhaps – if he was lucky. But, try as he might, he had totally failed to find any magic formula. Last time they had had sex, a week ago, just before her period, she had come. And he had done exactly what he had done tonight. He had got her excited during foreplay, almost to the point of coming, entered her quickly, thrust a few times, and she had come. Tonight, it had failed. Yet, sometimes, he could virtually miss out foreplay altogether, and she would come. At other times, if he tried to bypass foreplay, she would complain.

The woman lay on her side with her back to her partner, feeling disappointed. She hadn't really expected him to want sex tonight and now she wished he hadn't. As much as anything, she felt let down by the way, having made the overtures, he had misjudged what she wanted. There were times when she really wanted a penis inside her and wasn't particularly bothered about having an orgasm. And there were times when she really wanted both. Tonight, what she had really wanted was an orgasm. Surely he could have sensed that? She would have been quite happy to have intercourse as well, but what she really wanted was climax and relief – and she had been so nearly there. A few more seconds of foreplay and she would have come. Why couldn't he have waited just that bit longer before entering her? If he had just been con-siderate enough to concentrate on her until she had come, she would have been quite happy to have him inside her. But no, all he had wanted was to feel himself inside her, with no thought for her at all. She knew the second he stopped stimulating her and began to move into position that she would lose it. And by the time his penis had started to go in, she had. That fragile focus on her own sensations, which she needed if she was to climax, had simply evaporated. All excitement had gone and his interminable thrusting had killed it stone-dead.

She envied the women, and she was sure there were many, who needed only a few thrusts of a penis inside them to send them into multiple screaming, clawing climaxes. Sometimes she found herself looking at women in supermarkets, wondering what their climaxes were like. This morning there had been a short, dark-haired, wild-eyed woman in front of her at the check-out. As she had waited patiently for her own turn to pay she had had this image of the woman, naked on a bed, hair streaming about her shoulders, having energetic intercourse and screaming in ecstasy. For her, though, intercourse was almost always a disappointment. If she did climax during intercourse itself, which happened once a month if she was lucky, it was always a fairly tame affair; nowhere near as powerful as those during foreplay or those she gave herself.

The truth was, she wasn't particularly bothered about trying to come during intercourse. If it happened, it was a mildly pleasant bonus, but not really worth agonising and making a big effort over. Especially as most attempts were doomed to failure and mild recrimination. She would much prefer her partner to give her an orgasm when she felt like it, *then* satisfy himself inside her. What she needed was a partner who was sensitive enough to her needs to be able to judge what she wanted and when she wanted it. Should she really have to tell him every time? The man next to her really wasn't measuring up too well at the moment.

~

Women differ considerably in their response to intercourse. Some nearly always climax during intercourse; some never do. These differences are an important part of the overall picture of human sexuality, and we shall explore them in Scene 36. Taking women as a whole, it is more common to fail to climax with a penis in the vagina than to succeed. On average, only just over 60 per cent of routine sex episodes (from the beginning of foreplay to ejection of the flowback) involve the woman having an orgasm. Even when they do, she usually climaxes during foreplay (35 per cent) or post-play (15 per cent), not during intercourse itself. In fact, only 10–20 per cent of routine sex episodes involve the average woman climaxing while the penis is in her vagina.

There are several elements of interest in the scene we have just witnessed. The man's confusion is one, particularly his confusion over why an approach that stimulates his partner to orgasm on one occasion fails miserably on the next. Another is his observation that during his affair, his lover climaxed during intercourse much more easily than his partner. Yet another is why the man would have preferred his partner to climax during intercourse, not during foreplay. But these elements are the subjects of later scenes; for the moment, the main issue is the common lack of female climax during intercourse. Does it reflect some sort of failure? Or is it instead a successful part of a woman's subconscious strategy – yet another aspect of sexuality that actually increases her reproductive success?

When foreplay first begins, a woman's body will have carried out at least some advance preparation. The nature of this preparation has just been discussed (Scenes 22 and 23) in connection with nocturnal and masturbatory orgasms. Having these orgasms prepares her cervical filter in one way; not having them prepares it another way. Whichever option her body has taken, we know that she will settle down to intercourse with a certain level of protection against disease, a certain level of vaginal lubricant, a certain strength of cervical filter, and a certain population from nought to millions of sperm in her oviducts, womb and cervical crypts. If she has anticipated her situation correctly, these different levels will mean the state of her cervical filter is as ideal as possible, and she will get maximum benefit from the intercourse that is about to begin. Precisely what state of cervical filter is ideal varies according to her circumstances.

For example, the ideal filter will vary according to the stage of her menstrual cycle (Scenes 22 and 23). It will also vary according to whether or not she is contemplating infidelity (Scene 26). The ideal filter is different yet again if she is about to be inseminated by a man who is not her regular partner. In particular, it differs according to whether she wants the inseminator to have an advantage or disadvantage in any sperm war that might be about to take place (Scene 26). No matter what the ideal cervical filter may be for any given woman in any given circumstance, though, the general principle we are about to discuss remains the same. So in order to illustrate this principle, let's concentrate on the woman in Scene 24.

This woman had just finished menstruating. Ovulation (if it was going to occur this cycle – Scene 15) was therefore between seven and twenty days away, and the intercourse that had just finished would not make her pregnant. Nevertheless, the sperm from this intercourse could influence sperm retention at her next intercourse, which in its turn *could* make her pregnant (Scene 7). As far as we know, infidelity during her next fertile phase was unlikely – but, as always, was not impossible. So even though this current intercourse was during her infertile phase, it was important to her to retain an optimum number of sperm – for all she knew these sperm might suddenly, albeit indirectly, play a role in influencing the paternity of her next child.

Given her situation, all the woman needed from her intercourse was a small number of young sperm to store in her cervical crypts (Scene 4). Having such a reservoir would give her maximum flexibility over the next few days. When, sometime during those next few days, her body decided what cervical filter would be ideal for her next anticipated insemination, she would have all the raw materials necessary. (Without sperm in her cervical crypts, her options would be much more limited (Scene 22).) Moreover, the younger the sperm she could collect now, the longer her flexibility of choice would last. Of course, at the same time as collecting the sperm which would give her this flexibility, she needed as always to minimise the risk of disease.

Before intercourse, her cervical filter was strong because of the presence of menstrual debris (Scene 3). Most of the sperm from her last intercourse, a week earlier, had been carried from her body by her menstrual flow. A few infertile sperm might still have been in her oviducts and a few of her cervical crypts might still have contained a number of, albeit fairly geriatric, sperm. Although this set-up constituted an adequate filter, it would benefit from some 'tidying-up'. It is at this stage of the menstrual cycle that many women would either masturbate or have a nocturnal orgasm (Scenes 22 and 23), thereby helping to rid their cervical mucus of the last of the menstrual debris, top up their vaginal lubricant, and empty their cervical crypts of geriatric sperm. Such an orgasm would facilitate lubrication at their next intercourse. It would also, on balance, slightly strengthen their cervical filter – in particular, it would give them extra protection against invasion by any disease organisms carried in the next inseminate. But it would also minimise their retention of sperm, particularly old, less motile ones.

The woman in Scene 24 might have been due for a nocturnal, perhaps even later on in the night in question. If she had had one, it would have produced the ideal cervical filter for her circumstances. But an unexpected intercourse intervened before this could happen, and when her partner initiated foreplay, her cervical filter was not ideal for the inseminate she was about to receive. She could, of course, have resisted his sexual overtures and waited until her body had prepared her cervical filter properly. However, she did have another option, and this was the one she pursued. Having failed to prepare for this unexpected intercourse by having a nocturnal or masturbatory orgasm, her body instead tried to engineer an orgasm during foreplay.

A 'foreplay orgasm' performs all of the functions of a nocturnal or masturbatory orgasm, and for the woman in Scene 24 would have been a perfectly adequate last-minute substitute. The problem she faced in pursuing this option was that her strategy required cooperation from her partner. In part she needed him to provide stimulation. More importantly, she also needed him to give her time to climax. But he did not.

We shall discuss why he did not in Scene 25. The simple answer, is that he would benefit most from his partner climaxing during intercourse, not during foreplay. On this occasion, male and female interests were not mutual. This is often the case, which means that a woman cannot always rely on a man's cooperation when her body opts to have an orgasm during foreplay. The woman in Scene 24 nearly managed to engineer the foreplay orgasm her body needed to create the ideal scenario for insemination, but just failed. Having failed, her next best response was not to climax at all. And that is what happened.

When a woman has no orgasm during intercourse, then the number and types of sperm she retains are dictated by the cervical filter in place before the intercourse began (as is her resistance to invasion by disease organisms). If her body has anticipated events correctly, this filter will be just right to retain the best number and types of sperm for the next few days of her reproductive life. By *not* having an orgasm during intercourse, therefore, a woman's body is in effect saying: 'Don't change a thing. The situation in your cervix is as good as it can be. Just let him inseminate you and your cervical filter will do the rest.'

In the scene, the situation in the woman's cervix was not ideal when her partner ejaculated. When her best strategy was thwarted, her body, by avoiding an orgasm during intercourse, did the next best thing. Her failure to climax represented a success. (Precisely why will be made clear in the next scene.) We are about to watch the couple for just a little longer.

SCENE 25

Correcting Mistakes

They both found it difficult to relax. Their post-coital sleep was fitful, and after thirty minutes or so both were wide awake again. Suddenly, for no obvious reason, the man's penis began to stir and within minutes it was hard against his partner's back. There was no mistaking this feeling in his loins. He wanted sex again.

The woman had felt his penis rise and then harden, and wondered whether he would make any overtures. She found herself hoping he would, assuming her interest was because she had failed to climax during their recent foreplay. To her surprise, however, she realised that what she really felt like was having his penis inside her again. So clear were her feelings that when her partner hesitantly put his hand between her buttocks to stroke her genitals, she forestalled him by turning over to face him. After they had kissed a few times, she gently took hold of his penis and guided it into position so that he would be in no doubt as to what she wanted him to do.

As he slipped in, she was so wet from their previous intercourse he felt almost as if he wasn't touching the sides. When he first began his slow thrusting, it got worse. There was hardly any sensation on his penis except wetness. He was a long way from ejaculation.

She on the other hand, almost as soon as he entered her, felt the first sensation that told her she might come. As he worked backwards and forwards, backwards and forwards, she began to focus on her genitals, and it was working. Then came the fantasy. It was the dark-haired, wild-eyed woman from the supermarket queue. She was lying naked on top of her, black hair flying everywhere as she contorted her body, now to rub their genitals together, now to kiss her all over. With the combination of the sensations from her partner and her fantasy, the woman's body wound itself up towards a climax. The noises in her throat were becoming more urgent, more rhythmic. Just a little more focus, just a little longer . . .

By now, the man realised his partner was heading towards an orgasm. Her sounds were becoming more intense. He responded with his own sounds. The overwhelming wetness of her vagina had now gone, and he was beginning to get the sensations on his penis that he needed. But he still wasn't certain he was going to ejaculate. He worked on a fantasy. It wasn't his partner underneath him but the flirtatious seventeen-year-old from work. She was staring into his eyes and massaging her breasts with her hands. Between the guttural and rhythmic sounds from her throat she was telling him how wonderful he felt and how he mustn't stop. That did it, and only just in time – his partner was climaxing. As she did so, he briefly lost his focus, but still managed to ejaculate in the next half-minute.

For her, the climax was satisfying and just what she wanted. For him, their combined climax had given a sense of achievement and satisfaction. She didn't often come with him inside her and he wanted to make the most of the occasion. But he didn't. Almost within seconds of their expressing surprise and pleasure at the episode, they were both asleep.

~

At first glance, the couple in this scene have behaved very strangely. Why did they both feel like intercourse again so soon? Why did the woman now feel like intercourse and not fore-play? Why did she feel like an orgasm during intercourse now, when only half an hour earlier she had felt no such inclination? Could each person's behaviour really have been part of a strategy that over enough occasions would enhance their reproductive success?

As we know, a woman does not need to climax during inter-course for sperm to enter her cervix. Even without a climax, the inseminate collects at the top of her vagina to form a pool, her cervix hangs into the pool, and sperm leave to escape into her cervical mucus (Scene 3). Nevertheless, a climax does make a difference to *how many* sperm enter her cervix. Usually, a climax during intercourse weakens her cervical filter greatly, allowing many more sperm to leave the seminal pool, penetrate the cervical mucus, and hence be retained. Such a climax there-fore means fewer sperm are ejected in the flowback. A multiple orgasm (in which a woman has two – or even more – climaxes without 'coming down' in between) has an even greater effect on sperm retention. Essentially, this influence of the orgasm, whether single or multiple, is achieved in four ways.

First, when a woman climaxes during intercourse, her cervix gapes in just the same way as it does during masturbation (Scene 22). As we saw then, this gaping stretches the cervical mucus sideways, widening existing mucus channels and creating new ones by splitting the mucus, and thereby opening up many more pathways for many more sperm. Any blocked channels in her cervix are effectively bypassed and rendered ineffectual.

Secondly, as she climaxes during intercourse her cervix dips up and down as well as gapes – again, just as it does during masturbation. Now, however, it is dipping up and down in seminal fluid, thereby mixing the pool. This mixing helps more sperm, particularly older, less mobile ones, to contact and penetrate the cervical mucus.

Thirdly, the climactic contraction and rippling of muscles in the womb and vagina during orgasm generate pressure changes in the womb and cervix. These changes effectively 'suck' the usual fingers of semen (Scene 3) much further into the channels of cervical mucus. This stronger upsuck of semen into channels that are already increased in number and enlarged by the orgasm does two things. It helps to neutralise any acidity of the lower part of the mucus glacier, making it easier for sperm to escape the seminal pool into the cervix; and it increases the volume of semen in contact with the mucus, again allowing many more sperm to escape.

Fourthly, a woman's climax during intercourse voids many of her cervical crypts of old sperm. Although these ejected sperm may eventually block some of the newly formed channels in the same way as they do after a masturbatory orgasm, sperm newly arrived from the seminal pool still find that they have many more crypts all to themselves. Effectively, therefore, an intercourse orgasm creates more storage space for sperm to occupy.

This combined gaping, dipping and sucking of a woman's cervix during an intercourse orgasm permits her to retain far more sperm than if she had no orgasm during intercourse. As a rough guide, she retains 50–90 per cent of sperm if she has a climax during intercourse compared with 0–50 per cent if she does not. Indeed, so effective is a climax during intercourse at helping sperm out of the seminal pool that, no matter how strong the filter a woman prepares in advance, such a climax negates it.

In effect, then, when she feels like climaxing during intercourse, her body is saying to her: 'We have made a mistake in our preparation for this intercourse. Circumstances have changed, and our cervical filter is too strong. We need many more of the sperm that are on their way than we shall get if we don't do something. Bypass the filter. That way we shall let in more sperm and eject fewer in the flowback.'

Of course, for an orgasm during intercourse to have the consequences just described, there needs to be a seminal pool in the vagina. We should expect, therefore, that an orgasm during intercourse will be effective at bypassing the filter only if it occurs after the man has ejaculated. This turns out to be true, but the situation is not quite as simple as we might expect – because a woman's *subjective* climax during orgasm is not actually the critical event as far as sperm retention is concerned.

Even if she climaxes about a minute (two minutes at the most) *before* the man ejaculates, the orgasm will still permit large numbers of sperm to bypass her cervical filter. How can this be? It is because the moment she subjectively identifies to be her climax is only the beginning of a series of events in her womb and cervix which she cannot feel but which may continue for several minutes. The peak of the cervical activity that affects sperm retention actually occurs one to two minutes *after* her subjective climax. By the time her cervix reaches this peak, she is already beginning to relax. As long as the seminal pool is in place before her cervical activity peaks, her filter will be bypassed. In fact, as long as it is still in place, a climax even as long as an hour after ejaculation will allow her filter to be bypassed. There is no need for the man's penis to be in her vagina, and it makes no difference whether it is he who stimulates her to orgasm during this post-coital phase or whether she stimulates herself via masturbation. So even if a woman avoids climax while the penis is in her vagina, her body still has up to an hour to change its mind.

We can now discuss why, in Scenes 24 to 25, as the night progressed the woman's needs changed as far as orgasms were concerned. The word 'episode' will be used here to mean the whole time from the beginning of foreplay to the ejection of the flowback.

On the night in question we followed the woman and her partner through two sexual episodes. At the beginning of the first (Scene 24), her body's main requirement from intercourse at that stage of her menstrual cycle was to retain small numbers of sperm that were as young as possible. She also needed to minimise the risk of infection. To fulfil this requirement she needed to strengthen her cervical filter before her partner had any chance to inseminate. This would both improve her resistance to infection and make life difficult for the older, least motile sperm in his ejaculate. The right filter for her needs could have been produced a day or so in advance via a nocturnal or masturbatory orgasm, but her body had failed in this.

Even so, it was not too late. Her motivation and behaviour during the first episode were just what they needed to be, because although she failed to engineer an orgasm during foreplay, she did the next best thing and avoided climaxing during intercourse.

At the beginning of the second sexual episode (Scene 25), the woman's needs were still the same – small numbers of young sperm at minimum risk of disease – but the circumstances had changed. Most importantly, her partner was now offering to give her a small inseminate rich in very young sperm – younger than any in his previous inseminate. This was just what she had wanted all along, and it was this offer that made her body interested in a second insemination. However, the first insemination had left her with two problems.

First, her cervical filter was now stronger than before because of the influx of blockers from the first insemination. If she delegated sperm retention to her now over-strong cervical filter, she would lose more of the young sperm in the second inseminate than she would like. Secondly, she still had the seminal pool from the first insemination at the top of her vagina. If she had an orgasm during foreplay now, she would bypass her filter with large numbers of older sperm from the first inseminate – just what she had gone out of her way to avoid during the first episode.

Her solution to these two problems was straightforward. What she needed was to get rid of the first seminal pool, then to use an intercourse orgasm to bypass her cervical filter. That way she would retain a large proportion of the very young and desirable sperm from the second seminal pool. So how could she get rid of the old pool, given that it probably hadn't yet decoagulated enough for her to eject it as a flowback? Her best option was to encourage the man to remove it with his thrusting penis (as described in other circumstances in Scene 21) before introducing his second ejaculate. Her body generated the necessary urges. First, she no longer felt like foreplay, but wanted a penis inside her as soon as possible. Secondly, she felt like an orgasm during intercourse. Once intercourse had begun, her body then timed its build-up so that she didn't climax until the old seminal pool had been removed and her partner was nearly ready to ejaculate.

In this discussion we have, inevitably, considered only one specific sequence of events – the sequence relating to the woman in Scenes 24 and 25. And although we cannot, obviously, discuss every combination of circumstances and responses, there are some general principles worth noting.

First, a woman prepares for her next anticipated intercourse by having a nocturnal or masturbatory orgasm – or by *not* having a nocturnal or masturbatory orgasm – whichever is appropriate, as explained above. Secondly, if she has anticipated correctly, she should let the entire sexual episode proceed without orgasm and leave her filter to do its job. On the other hand, if she has not anticipated correctly, she should aim to correct her mistake by having an orgasm at some time during the episode.

In the latter situation she has two options, depending on what type of mistake she has made. The fact that orgasms during foreplay are more frequent than orgasms during intercourse suggests that a woman is more likely to err on the side of having too weak a filter than too strong. This could be strategic. It is easier to encourage a man to help her climax during foreplay (strengthening her filter) than to help her climax during intercourse (weakening her filter). Once she allows him to insert his penis into her vagina, she loses much of her control over what he does and when he does it. Even then, however, she has a fall-back strategy: orgasms after the man has withdrawn are a substitute for intercourse orgasms, permitting her to bypass her filter whenever she needs or wants to (again, as just described).

The best moment for a woman to climax (or not) during a sexual episode thus varies considerably from occasion to occasion. The best moment for her to climax as far as the man is concerned, though, varies much less. As we have already seen (Scenes 12 and 14), he also will have anticipated the intercourse and tailored his ejaculate accordingly. He will have masturbated or not, as appropriate, to suit his body's own best interests. Most often he will be best served by a woman climaxing during intercourse, because only by this means will his carefully prepared inseminate succeed in having as many of its sperm retained as possible.

But there is a limit to how much time and effort it is worth him putting into getting a woman to climax during intercourse. First, if her body is not interested in an intercourse orgasm, he cannot force her to climax, no matter how hard he tries. Secondly, the cervical sperm filter which he can sometimes try so hard to bypass may on some occasions not actually exist. *If* the woman has retained very few sperm from her previous insemination, *if* she has neither masturbated nor had a nocturnal since that insemination, and *if* any even older sperm or menstrual debris have been cleared, the filter in her cervix will be as weak as it can be anyway. In this case, the man gains little if anything from making a great effort to encourage her to climax during intercourse.

All of this leaves him with a very difficult, if not impossible, series of decisions to make during sex. First, he has to judge whether the woman has a filter that is worth bypassing. If not, he has no need for her to climax during intercourse. If she has a significant filter, his best option is for her to have an orgasm during their sexual episode. Even an orgasm during foreplay will do, as long as he manages to inseminate her within the minute or so afterwards. If he opts instead to try to get her to climax during intercourse, he may have trouble succeeding, and if so he will have to decide how long it is worth carrying on thrusting. He could opt instead to ejaculate first and then see if he can stimulate her to climax afterwards, before she ejects the flowback. Or he can just abandon all attempts to bypass her filter and hope that this is an occasion when there is no real filter to bypass.

Looked at in this way, a sexual episode is actually a contest between the man and the woman. Except on those rare occasions when the same outcome, bypassing the filter, suits them both, each is forever trying to manoeuvre the other into doing something that is against their body's interests. Whether the man succeeds in what is his safest ploy – to ejaculate a few seconds after the woman has climaxed – inevitably depends more on the woman than on himself.

A woman may insist on a foreplay orgasm, then cooperate or hinder as the man tries to inseminate her within a minute or so afterwards. If she curtails foreplay, several strategies are possible during intercourse itself. She may cooperate by waiting until he is about to ejaculate, then climaxing with him. Or she may not cooperate. She may climax so quickly that he is unable to ejaculate within the critical minute that follows. Or she may avoid climaxing and wait for him to give up and just go ahead and ejaculate – in which case intercourse becomes a war of attrition, each waiting for the other to climax. Once the man *has* ejaculated, the woman can either avoid climaxing before ejecting the flowback, or she can cooperate by helping him to give her a post-coital climax. Or, almost the ultimate cooperation, she can 'finish herself off' by masturbating to orgasm before ejecting the flowback.

It might, at first, seem strange that, given the competitive nature of intercourse, men and women appear to make it easier for each other to judge how near they each are to their climax. Even to the objective observer, the sounds made by men and women during intercourse give a clear indication of how their levels of sexual excitement are changing (which is why the more exaggerated of such sounds are dubbed on to pornographic films). When climax comes, it is clear from the sounds, which would seem to indicate a cooperation between a man and woman that might belie our interpretation of intercourse as a contest. The sounds, however, have become part of the contest.

Sometimes, of course, the sounds really do give an accurate indication of what is happening. On those occasions when a virtually simultaneous climax is in both people's interests, they are the method used to achieve this mutual aim. But it is precisely because the sounds are sometimes honest that on occasion both men and women can use them to trick the partner. In surveys, over half of women admit to faking orgasms sometimes, and a quarter to doing so often. That they often succeed in fooling the partner is evident from another survey – which found that the man invariably reports a higher frequency of simultaneous orgasms than his partner. There are cases on record in which a man has reported that his partner always climaxes during intercourse whereas she herself has stated that she never does.

The scene we have just witnessed was one of those relatively infrequent occasions when both the man and the woman had the same requirements from their intercourse, and cooperation was in their mutual interest. In the next scene, however, the needs of the woman we meet are very different from those of one of the men who inseminate her. The female orgasm is about to enter the arena of sperm warfare.

SCENE 26

Putting It All Together

The woman lay back in her Friday-night bath and let the last waves of her orgasm subside. She had finished her period only a week ago and this was the second time since then. Both times the thought and the action had come out of the blue. A minute earlier she had been soaping and rinsing her breasts, with nothing more erotic on her mind than whether she should clean the bathroom this weekend or suggest her partner do it while she was away. Maybe it had been the touch of her soapy hand on her nipple, but the next second the thought was in her mind, the tingle was between her legs, and she knew she needed relief. Now, after five minutes of successful cooperation between her fantasies, her finger and her clitoris, her breathing was easing, her heart was slowing down, and she was wallowing in the combined warmth of bath water and post-climactic glow.

To masturbate twice in five days was unusual for her; once a week or once a fortnight was more like it. She wondered if it was because she had given up the pill recently, having convinced her partner that she really wanted another child. However, she did not question her body too closely – she was enjoying this atypical burst of sexuality. She mused over the way that two nights ago, only a couple of hours after she had masturbated, she had then more or less insisted that her partner give her an orgasm during foreplay as well, gently resisting his attempts to penetrate until he complied. Three orgasms in a week. She was doing well.

Over the following weekend, she had routine sex with her partner only once, on Saturday, and again he obliged her with an orgasm during foreplay. Early Sunday morning, just before she woke, she had a nocturnal, her fifth orgasm that week. However, it was the last exciting thing to happen that day. The remainder was spent travelling, on the long drive to and from her parents, who had volunteered to look after her two daughters while she was away. By the time she and her partner had returned home and she had packed, it was late. He still wanted to have sex. 'One for the road,' as he put it. But she was tired and managed to put him off. On Monday, she left for the airport and her long-awaited three-day conference trip abroad. Her companions were to be her boss and two other colleagues, one of whom was her new lover.

Actually, 'lover' was an exaggeration, or at least premature. They had known each other for nearly a year, ever since she had joined the company. It was her first job since the birth of her younger daughter. Slowly, they had grown close. Neither had ever found it easy to make friends with their own sex, but in each other they had found the nearest they were likely to get to a 'best friend'. In their early thirties, both had young children and both were now in long-standing but increasingly stale relationships. He was even convinced that his partner was having an affair. Within a few months of meeting, they had freely acknowledged their physical attraction to each other. He had tried somewhat half-heartedly to do something about it, but each time he suggested a plan, she had found some excuse. It hadn't been difficult.

They had also acknowledged that they could never live together. They were both opinionated and competitive, with strong likes and dislikes. Their relationship worked only as long as they talked about their separate problems, their separate aims, and their separate wishes. Whenever they talked about principles, attitudes or just life in general, they would argue. She could not imagine a home without at least three cats; he hated cats. She was pale and prone to sunburn; he was swarthy and worshipped the sun, often spending hours naked in his secluded garden. She was aggressively vegetarian; he could not imagine life without meat. She was obsessively tidy; he still had the air of an adolescent, putting his feet on tables and dropping his crushed beer cans on the floor. She was provocative and mischievous, a tease; he was gullible, but strong-willed, ambitious and obstinate. They each found the other's traits to be irritating but attractive, each seeing in the other characteristics which moderated their own polarity.

Their relationship had changed six weeks ago. On that day, in a field in searing sun, she had given him oral sex and promised that one day she would have his baby. True, she had done so in part as a ploy to avoid having sex with him there and then, but she had meant it. She just hadn't expected the opportunity to arise quite so soon. Then, as early as two weeks after her promise, they had been asked by their boss to accompany him on a business trip abroad. Her future lover had then made her promise that if they had the chance, they would have sex while they were away.

Even so, she still didn't give in easily. He visited her hotel room almost as soon as they arrived early on the Monday afternoon, but she sidestepped his advances, pointing out that this was their only chance to see the city during the day. Together, they mingled with summer tourists in the streets and on the canals and enjoyed each other's company, watching the world parade past as they sat and sipped at pavement bars. By late afternoon, they felt extremely close as they arrived back at the hotel. But still she resisted sex, saying she only just had time to get ready before they met up with their boss for the evening.

The meal and the socialising were a strain because they had to maintain a façade of distance. They had both agreed that nobody must suspect they were so close. It was midnight when they arrived back at the hotel. She pointed out that the next day would be very busy. She tried to persuade him that they needed a good night's sleep and that maybe they should wait until tomorrow night. But this time, he would not be deterred.

She had seen him naked before, but it was the first time he had seen or really touched any of the more exciting parts of her body. By the time they were both naked, because still she teased and delayed, he was on the verge of ejaculation. With scarcely any foreplay, he tried to enter her. She objected strongly to his urgency, but he was now so desperate that he virtually forced himself inside her. He ejaculated immediately, with scarcely a thrust.

Once he had withdrawn, she chastised him. He apologised and said that next time he would be more considerate. She said there wouldn't be a next time if he behaved like that. But she didn't throw him out of her bed and within fifteen minutes he had stroked, kissed, licked and massaged her into a climax. Within the hour, his penis was back inside her for the second time that night, and after a long and sensuous intercourse they climaxed more or less together.

Half-way through the night, as a sperm war was being fought inside her body, she told him to go back to his room. When he protested, she said she was worried about them being discovered together in the morning. Over the next two days, they maintained a professional distance, both during the day and while socialising in the evenings. But on both nights they spent a few hours in bed together before she made him return to his room. By the time they left to fly home they had had intercourse six times, and on three of those occasions she had climaxed while he was inside her. Her body was full of her lover's sperm, from cervix to oviduct – and she was only forty-eight hours from ovulation.

She arrived home late that Thursday night, feeling very tired, very guilty and a little nervous. She bathed, avoided her partner's advances by pleading tiredness and nausea, and slept. Sometime during the night, she had a nocturnal.

At work, on Friday, she tried to avoid her lover as much as possible. He sought her out at every opportunity and tested her ingenuity to its limit. Eventually, when he tried to snatch a kiss, she rounded on him. Their trip had been fantastic, she told him, but now they were back things had to return to normal. She didn't want to hurt her partner and she didn't want to risk discovery. They had had a good time and it seemed they had got away with it. She wanted it to stay that way. If he wanted to remain close to her, he had to forget sex. A friend she needed, a lover she didn't. She felt guilty at the look of hurt on his face as she spoke, but her tirade worked. After that, he made no further sexual advances.

That evening, the last she and her partner were to have on their own before collecting their daughters, she bathed, masturbated, and then appeared naked in their lounge urging her partner to give her another baby. They had intercourse on the floor, just the once, and she pretended to climax with him. That pleased him because these days she rarely climaxed with him inside her. They then sat and relaxed in the warmth of the summer's evening, a cat on each lap. As she gave him the public version of her business trip, sperm warfare resumed inside her body and continued all night. But this war was very one-sided.

As they drove to her parents' the next day, a sperm entered the egg which had just arrived in one of her oviducts. She had conceived, and with her help her lover's sperm had won both battles, and the war. She had kept her promise to him.

∾

This, of course, is the third time we have met this woman and her lover. We watched them in a hay-field on a sunny afternoon as she gave him oral sex and promised one day to have his baby (Scene 20). We also followed them that same evening as they returned home to their respective partners and had oral sex (Scene 10). Now, we have just watched the woman's body manipulate sperm warfare in a way that made sure she kept her promise. Her third child would inherit her lover's looks, drive and ability in combination with her own cunning and intelligence. Her body had decided that as the genetic father of her third child, her lover was the best compromise around (Scene 18).

In Scenes 22 to 25 we have unravelled, step by step, the function of the female orgasm. We have seen the effects of having a masturbatory or nocturnal orgasm in the days before intercourse, and of not having one; of climaxing, and of not climaxing, while a seminal pool is at the top of the vagina. In this most recent scene we witnessed a woman's body putting all of these orgasms together and biasing sperm warfare in her own best interests. As a result, she conceived a child who was fathered by her lover, not her partner. This move will have enhanced her reproductive success – as long as her body was correct in judging that a child by her lover would be more successful reproductively than a further child by her partner.

Having reached its decision over the best father for its next child, her body set about engineering paternity. It did so by generating and orchestrating a sequence of urges to climax and not to climax. Her body dictated the sequence and timing of these urges according to who was the next most likely inseminator. We can now discuss the details of this and other such sequences. Of all of the weapons used by a woman in her pursuit of reproductive success, her orgasm sequence may well be the most important, particularly when she promotes sperm warfare.

This is the second scene in which a woman has increased her masturbation rate in the context of infidelity. In Scene 6, the woman masturbated more often than usual in the week before infidelity with her ex-boyfriend. At first glance, such events seem simply to reflect an increase in sexual excitement – the anticipation of sex with someone other than a partner. Closer inspection, however, shows that this is too simple an explanation. The women in the two scenes masturbated or had nocturnals *only* at times when, although infidelity was on the horizon, the person most likely to inseminate them next was their partner. On no occasion in these two scenes did they have these orgasms when their very next intercourse was likely to be with their lover. In this, women differ from men (Scene 13).

Of course, women do have nocturnal and masturbatory orgasms just before having sex with a lover, but these are often 'mistakes'. In Scene 26, the woman had a nocturnal on the Saturday night, yet as it turned out her next insemination was by her lover on Monday night. At the time she had her nocturnal, though the next *most likely* intercourse was the 'one for the road' suggested by her partner on Sunday night. As it happened, this insemination never materialised. Her nocturnal was a 'mistake', but one she could later correct, as we shall see.

So what is this link between infidelity and the frequency of nocturnal and masturbatory orgasms if it is not simply sexual excitement? And why, if they are a response to the anticipation of infidelity, are these orgasms more likely to occur just before sex with the partner rather than just before sex with the lover?

It is because whenever a woman anticipates sperm warfare, she more often than not has some preference over who should win. Usually, that preference is for the lover, not the partner – otherwise, she would not risk all the potential costs of infidelity (Scenes 8 to 11). In this scene, just as in Scene 6, we have watched a woman who, in anticipation of sperm warfare, subconsciously prepared the battlefield to favour her lover's army.

In the current scene, the woman's body had decided that the lover would be a better genetic father to her child than her current partner, but a worse long-term partner. Her body's ploy, therefore, was to collect sperm for fertilisation from the lover, but not to do anything else that might cause her to lose her current partner. This meant she would have to have sex with her partner both before and after her infidelity. Otherwise, his suspicions would be aroused. Consequently, she could not avoid sperm warfare. In any case, her body would not want to miss out on the benefits of promoting such warfare (Scenes 17 and 21). Her best option, therefore, was to make life as easy as possible for her lover's army while still giving it a contest to win. As events turned out, her body engineered the situation perfectly.

The woman's body initiated its strategy a week before the anticipated infidelity. Its aim was to allow routine sex with her partner to continue undisturbed, while ensuring that his army would be as small as possible when the troops were eventually called to battle. Her body's technique was first to make her feel like masturbating a day or so before each anticipated routine intercourse, and this strengthened her cervical filter. Then, during each routine sexual episode, it made her feel like an orgasm during foreplay. This orgasm strengthened her filter still further.

In the event, the woman and her partner had two episodes of routine sex during the critical week before her infidelity. On neither occasion did she feel like an orgasm during intercourse itself. So the result of her body's preparations was that on neither occasion did she retain many of her partner's sperm. Her nocturnal on the Saturday night was a continuation of her body's strategy. There was a strong possibility that her partner would want sex on the Sunday night, just before she left on her trip. The nocturnal prepared a strong filter, ensuring that even if he had insisted, she would still have managed to keep down the number of his sperm in her tract – especially if she also had yet another foreplay orgasm during the episode. As it was, she managed to avoid insemination by her partner on the Sunday night altogether.

When the woman was eventually inseminated by her lover on the Monday night, she contained only a small army of her partner's sperm, just as her body had planned. However, she still had a reasonably powerful cervical filter, not least because of the unnecessary nocturnal on the Saturday night. She tried to delay the first intercourse with her lover for another day to give her partner's army time to decrease in size still further, but her lover would not be delayed any longer.

Because of the strength of her filter, when she did have sex with her lover for the first time, her body felt like an orgasm during intercourse. That way, she could correct the 'mistake' of her earlier nocturnal preparation and bypass her filter. Her lover, though, was pursuing a strategy of his own. Having been thwarted in his attempts to have sex with her on so many previous occasions, his absolute priority was to ejaculate inside her. When at last she allowed penetration, he ejaculated as quickly as he possibly could, before she changed her mind yet again and made him withdraw.

His urgency prevented her body from bypassing her strong filter during intercourse, and at first few of his sperm managed to escape from the seminal pool into the cervix. But, with his cooperation, she climaxed from his caresses fifteen minutes later. With the seminal pool still in position, his huge army now bypassed her filter and flooded her cervix and womb. The lover had got the best of both worlds – rapid insemination before she changed her mind, and a high level of sperm retention. And she had satisfied her body's urge to collect as many sperm from this man as possible. As long as his army was reasonably competent, he should now easily defeat the remnants of her partner's small and ageing army.

An hour later, when the woman's lover offered to top her up with a small army of young sperm – his second inseminate – she accepted and they cooperated to produce an intercourse orgasm. This bypassed the filter in her cervix – although that filter was now substantial, containing blockers from both men – with the result that a high proportion of the lover's young small army reached her cervix and womb. The combination of his first relatively huge army and his second small, young army had an easy victory while she later slept.

Over the next two days, he kept her topped up with young sperm – and on half of the occasions she bypassed her filter with an intercourse orgasm. On her return home, she contained a large and powerful army from her lover. Her body's aim was now to engineer a strengthening of her filter before allowing her partner to inseminate her. She succeeded. First, she avoided sex with him any sooner than she had to without arousing his suspicions. Secondly, she had a nocturnal. Thirdly, unable to rely on his cooperation over a foreplay orgasm after such an absence, she took the safer option the next evening and masturbated – then immediately solicited sex.

This was to be the insemination that deceived her partner into thinking he could be the father of any child conceived that month. She felt no urge to climax during intercourse, only an urge to pretend that she did. As a result, few of her partner's sperm made it through her cervix into the battlefield. And those that did would be out-competed by the numerical superiority of her lover's killers and egg-getters. There was never any doubt that this particular war would be won by the woman's lover.

Her strategy worked very successfully, but of course none of it was achieved at a conscious level. Her body achieved its aims by subconsciously orchestrating the sequence of moods, motivations and responses that would best serve her reproductive interests. Consciously, she will simply have relished her sexual pleasures, her excitements and her fears as she cleverly negotiated the slalom of infidelity and deception.

The behaviour and responses of this woman during her infidelity were fairly typical. In Britain, the average frequency of nocturnal and masturbatory orgasms for women during phases of *fidelity* is just under one per week (Scenes 22 and 23). During phases of *infidelity*, the rate increases to about every other day. These orgasms occur most often when women anticipate that their next intercourse will be with their partner. When they anticipate it will be with their lover, they occur less often.

Women also change their likelihood of having an intercourse orgasm during infidelity. On average, they are more likely to climax during or after intercourse with a lover (33 per cent of occasions) than they are during or after routine intercourse with their partner (22 per cent of occasions). Thus a lover's army receives assistance in entering the battlefield more often than does a partner's.

These differences mean that on the whole not only does a partner encounter a stronger filter than a lover, he is also less likely to receive help in bypassing that filter. On average, the advantage enjoyed by a lover in sperm warfare is relatively large. When a woman *is not* being unfaithful, she helps her partner to place a large sperm army inside her on 55 per cent of occasions. When she *is* being unfaithful, she helps him on only 38 per cent of occasions, but helps her lover on 65 per cent of occasions – nearly twice as often.

The really impressive part of a woman's strategy, however, is the way she manages to engineer a strong bias over sperm retention in favour of her lover without betraying any hint of infidelity to her partner. First, she keeps the frequency with which she climaxes during or after intercourse with her partner the same (22 per cent of intercourses), whether she is being unfaithful or not. Secondly, her main weapon against her partner is to increase her frequency of masturbation and nocturnals, but because these orgasms are secret, the change passes undetected by him. Thirdly, her main weapon to favour her lover is to have fewer nocturnal and masturbatory orgasms before, and more bypass orgasms during, sex with her lover than with her partner. Yet again, her favouritism cannot be detected by her partner.

We can now see why it is as important for a woman to hide her masturbations and nocturnals from her partner as it is for a man (Scenes 12 to 14). If a man knew exactly when and how often his partner masturbated or had nocturnals during their routine sexual activity, any change in pattern could alert him to the fact that her body might be anticipating infidelity. With this warning, he could then be more vigilant, guard her more intently, and generally make it more difficult for her to collect anybody else's sperm. Her strategy for influencing the outcome of sperm warfare, therefore, just like a man's, depends on being able to change her masturbation and nocturnal patterns without being detected. Female masturbation and nocturnals have therefore been shaped to be secretive just as rigorously as male masturbation and wet dreams. And again, just as for men, it is the involvement of masturbation and nocturnals in infidelity and sperm warfare that has *generated* in women the subconscious urge for secrecy and privacy. The fact that, by and large, people do not know each other's masturbation patterns is a testament to the success of that urge. For all the same reasons that we discussed for men (Scenes 12 to 14), the urge for privacy and secrecy over masturbation shown by most women is often matched by their curiosity, suspicion, distaste, and even prejudice regarding the practice in general.

Of course, women do not always climax in secret. Just like a man, a woman will sometimes do so in full view of her partner without them then carrying on to have intercourse (Scene 20). Sometimes she does so by an open display of masturbation, but most often she does so with his help. We have already discussed facets of this situation with respect to oral sex (Scene 10), but often a man stimulates the woman with only his fingers. Here it is the orgasm itself that is strategically important to both the man and the woman – not any information the woman wishes to give or the man wishes to collect by his smelling and licking of her genitals.

Such open orgasms, without intercourse immediately following, have exactly the same consequences for sperm retention as a woman's private masturbation and nocturnals. Studies show that all three have an equally strengthening effect on her cervical filter, particularly if at orgasm she still has a large population of sperm in her cervical crypts. The result is that all three reduce sperm retention at her next insemination, even if this does not happen until days later.

For the man, there is an element of reassurance in seeing or helping his partner have such an orgasm. His subconscious mind registers that, if she does have sex with another man over the next day or so (for example, through infidelity or rape), at least she starts that intercourse with a strong filter, and any army placed in her tract will be reduced. Of course, the man's ploy is not infallible, and his partner can render his preparation ineffectual by having a bypass orgasm with the other man. Moreover, if, as is most likely, her next intercourse is with her partner, his preparation will have been counter-productive – he will have strengthened her filter against himself.

As we have noted, helping a woman to climax during foreplay is, on the whole, disadvantageous to the man. Unlike the climaxes between sexual episodes that we have just discussed it is *always* him, never another male, who is going to suffer lower sperm retention. The other disadvantage here is that the longer he waits to penetrate the greater the chance he will lose the opportunity – either the couple could be disturbed, or she might change her mind about allowing intercourse.

It is probably for these reasons that, as in many scenes in this book, most men would prefer to enter a woman and begin intercourse without trying to help her climax during foreplay. But because women often have much to gain from a foreplay orgasm (Scenes 24 to 25) and frequently seek a man's cooperation in achieving one, the level and length of stimulation during foreplay is one of the major areas of conflict between men and women during their sexual encounters.

A man is most prepared to cooperate over foreplay orgasms when he has least to lose, that is, when the woman either has no sperm in her cervical crypts or when the inseminate he is about to produce has little chance of becoming involved in sperm warfare; and when he has least choice, that is, when the woman refuses to allow penetration until she has had an orgasm during foreplay. So it is not surprising that men are most likely to cooperate over a foreplay orgasm when they have spent most time with their partner (so infidelity is unlikely), and when the woman is most insistent. Even having helped the woman to climax during foreplay, however, a man often shows great urgency in his attempts to penetrate and inseminate – because if he succeeds within one to two minutes of the woman's climax he can still bypass her cervical filter (Scene 25).

As we have seen time and time again, most of the strategies shown by men and women in relation to ejaculation and orgasm are subconscious – orchestrated by the body via sequences of mood, libido and sensitivity to stimulation. Indeed, most of the behaviour described in this book is similarly subconscious, the product of genetic programming rather than cerebral rationalisation. But the conscious element features none the less, as men and women learn by trial and error how best to satisfy their feelings. Men have to learn many things, from the basics of penetration to the subtleties of the female orgasm. Women have to learn how to climax, how to encourage men to help them to do so – and when and how to fake it. Both sexes also have to learn the strategic subtleties of infidelity and the prevention of infidelity; and how to select a mate, how to court and seduce the selected mate, as well as how to avoid unwanted attention.

The ability to learn all of these things both well and quickly, making as few mistakes en route as possible, will have a big influence on a person's reproductive success. Scenes 27 to 29 are concerned with how men and women learn these necessary sexual subtleties.

9

Learning the Gropes

SCENE 27

Practice Makes Quite Good

As the young man rolled off the girl beneath him on the bed, a pile of coats and sweaters collapsed on top of him in the dark. He pushed them away, on to the floor. He had done it. At long last, he had done it. Without a shadow of a doubt, he had finally ejaculated inside a woman.

He had come close twice before. The first time, when he was sixteen, was during a heavy petting session with a girl two years younger than himself. On that occasion, he had been content just to ejaculate, and hadn't really tried to enter her. Then, last year, at a party just like this one, he had tried again but failed. Convinced that he was inside the girl, he had happily thrust away until he ejaculated. It wasn't until afterwards that she had told him that all he had done was thrust down between her buttocks and ejaculate on to the bed, or maybe somebody's coat. This time, however, he had definitely succeeded. Nineteen years old, and he had lost his virginity.

Admittedly it had been quick; at the most a few seconds, once he was inside her. Admittedly, also, he had needed help. As on the previous occasion, the prods of his penis had failed to find the girl's vagina – though he wouldn't have known. Once again, he had thought he was inside her and had begun thrusting. But when she had taken hold of his penis and guided him in, the sensation was so different that he realised he must have been somewhere else – again, probably between her buttocks. He had nearly ejaculated prematurely, but had hung on just long enough to push himself right in. As he lay in the dark by her side, his sense of achievement and satisfaction knew no bounds. He told her how fantastic it had been and asked how it had been for her.

She said, 'Great' with a sarcasm that would have deflated anybody but a first-timer. When she had first seen him at the party, she had dismissed him as a prospect, thinking how young and inexperienced he seemed. But when he homed in on her, she had changed her mind. He was quite good-looking and his clothes were expensive. His conversation had been a little immature and naïve, but his witty stories of achievement and failure had impressed her. When he had brazenly asked her if she felt like having sex with him, she had been just drunk and randy enough to think it might be a good idea. As soon as they were in the dark room, he had more or less ripped off her underwear. Then, he had missed out all foreplay, lain on her like a dead weight, and prodded between her buttocks and the bed. Even when she helped him inside her, he had ejaculated as soon as he was in. Now he wanted praise. She felt embarrassed, annoyed and sexually frustrated; she vowed that, next time, she would take more notice of her first impressions.

Someone knocked on the bedroom door and, getting no response, tried to push it open. When the bolt on the inside held, the prospective intruder urged whoever was inside to hurry up, reminding them that they weren't the only ones who needed the room. When the girl began to shuffle around on the bed, the young man asked her what she was doing, saying they needn't give up the room just yet. It was so dark that neither could really see the other. The girl said she was looking for her knickers, and could he remember what he had done with them? He had put them on the floor somewhere, he said, and would look for them in a minute; there was no rush he repeated. Embarrassed that she had really allowed such a nonentity to have sex with her and eager to put the experience out of her mind, the girl lied that she needed to go to the toilet, and would he please help her to look? Reluctantly, he pulled his pants and trousers up from around his knees, fastened his zip, and got off the bed to help her.

In truth, he couldn't remember what he had done with her knickers. He had been astounded that this girl, whoever she was, had agreed to come into the bedroom in the first place. He had only talked and danced with her for about an hour. Even once he had got her into the bedroom, he was so certain that she would change her mind any second that he had taken off her tights and knickers with the speed of light, while she was still compliant. In his hurry, he could have thrown her underwear anywhere.

He asked what colour her knickers were. When she said they were black, it seemed clear that there was no chance of finding them amongst all the coats and sweaters littering the room. He suggested they switch on the light, but she told him to forget it, she would go without. The next second she was at the door, fumbling with the bolt. As the young man stumbled over the clothes towards her, the door opened and she was gone. No sooner had she left than the next couple were on their way in and he had to push past them. By the time he was out of the door, the girl had escaped down the stairs and immersed herself into a conversation.

Partly at first to avoid any further attention from the youth she had just escaped, she attached herself to the oldest man at the party. He was nearly thirty, ten years older than her. She knew who he was and knew also his reputation as a womaniser. Good-looking and reasonably successful, he told her he had managed to come to the party only because his partner had gone to visit her mother for the weekend. She stayed with him for the rest of the evening and became totally captivated by his charm and gentleness, by his humour and sensuality. When he suggested, as people began to leave, that she should let him drive her home, she accepted. When she suggested, as they kissed in his car and his hand discovered she wasn't wearing any underwear, that he should let her share his bed for the night, he accepted. She spent the rest of the night, all the next morning and part of the afternoon being kissed and stroked, pampered and stimulated. Three times he brought her to orgasm and four times he inseminated her. In between their love-makings she slept and dreamed of a long relationship with him.

For a while after that weekend, she managed to live out her dream. She remained the man's lover until she discovered that, as well as his partner, he was having sex with at least one other girl besides herself. Then she stopped meeting him.

At about the same time, her flatmate started seeing the young man to whom she had lost her knickers on the night of the party. After the pair had been an item for a few weeks, the girl's curiosity got the better of her and she asked her flatmate what he was like in bed. She could believe her friend when she said that at first he had known nothing. Evidently, however, he was learning fast, and a few days earlier he had actually found her clitoris. All he had to do now was learn what to do with it – then, she thought, he might become quite good. She had high hopes.

～

Sexual technique does not come instinctively to a man. It has to be learned. In this respect he is no different from all male birds and mammals. Arousal, erection and ejaculation are pre-programmed and automatic, but the niceties of sex have to be acquired. If a male is ever to persuade a female to allow insemi-nation, he has to pick up the subtleties of courtship and stimulation. He then has to learn how to copulate quickly and efficiently so as not to miss the opportunities his courtship technique has provided.

Male birds, for example, first have to learn how to stand on a female's back, then how to bend their tails to press against her genitalia before transferring sperm. Male mammals have to learn what to do with their erections – where to put their penises. Even as intelligent an animal as an adult male chimpanzee is totally inept if denied sexual opportunity during adolescence. His sexual development depends first on watching others having sex, and secondly on practising intercourse himself. In the company of a female, an inexperienced adult male such as this becomes aroused and erect but has no idea what else to do. He even has trouble knowing which end of the female to approach with his penis, and rarely succeeds in copulating on his first encounter or even on the next few. Thus, if any male mammal is not to miss the first of his lifetime opportunities to inseminate a female, he needs to practise during adolescence – and human males are no exception. As the youth in the scene discovered to his cost, failure to learn sexual technique quickly during adolescence can soon result in missed opportunities for insemination. And it can have an important influence on a man's level of reproductive success.

In all human cultures, young males first hear and see the basics of sexual technique from those with experience, either precocious peers or older educators. In many cultures, sexual experimentation by young, even pre-pubertal, boys and girls is openly encouraged or at least condoned. A boy who learns early how to persuade a girl to have sex with him, how to prepare her genitalia for penetration, how to stimulate her to lubricate, and how to find and enter her vagina with his erect penis will be less likely to miss his first opportunities to reproduce. The youth in Scene 27 had his first chance to learn sexual technique at sixteen years of age. This single occasion was not enough. Two years later he missed his first opportunity to inseminate a girl because he had not yet learned either how to find her vagina with his penis or what it felt like once he was inside. At nineteen, albeit with help, he managed to find a girl's vagina and inseminate her, but even then his inexperience cost him a full relationship with his conquest and, therefore, any further sexual opportunities with her.

Of course, a man has to learn much more than simply how

to gain a woman's sexual interest, how to retain that interest long enough to get the opportunity for intercourse, and where to put his penis when the moment finally arrives. A man who can also learn how to influence a woman's orgasm pattern has much more chance of influencing her pattern of sperm retention (Scenes 24 to 26). But a woman's body does little to make it easy for males to learn the necessary techniques. In fact, it does the converse. Why? The answer lies in mate selection, and the way in which women collect information about men.

We have already discussed at some length the criteria by which a woman selects a man (or men) to be her short- or long-term partner (Scenes 18 to 21). There we considered the importance of a man's status, behaviour, looks, fertility and sexual health. We also discussed how most mate selection is a process of compromise; how a man who can seem the best bet at one stage in a relationship may not seem so later; and how a woman sometimes needs to balance a man's visible qualities, such as his status and looks, against his less visible qualities, such as his prowess in sperm warfare.

In order to collect information about a man, a woman in effect needs to set him a series of tests. Depending on how many tests he passes compared with other available candidates, she will then either accept or reject him. She needs to set tests that are challenging but not impossible. They are of no value to her if they are too easy, or if they are so difficult that no man can pass them. A woman's body and behaviour have been shaped to present such a test. And as often as not the male quality being tested is his ability to learn how to use her body and cope with her behaviour.

Learning is always more difficult when a given behaviour does not always elicit the same response. The female response to any particular male stimulus is notoriously unpredictable. This is true from the earliest moments of courtship to the occurrence of orgasm during intercourse. Not only do women differ (for good reason – Scene 36), but individuals differ from occasion to occasion (again for good reason – Scenes 24 and 25).

This variability allows the female to set challenging, but passable, tests for those males who satisfy their other mate-selection criteria. Inevitably, these tests are most difficult for inexperienced males. As an example, let us consider the position of the clitoris. As we have already noted in humans, apes and many monkeys this organ is small, difficult to find, and removed from direct stimulation by the penis during intercourse (Scene 22). It *can* be stimulated by some part of the male's anatomy during intercourse – usually the penis – but unless the male knows *exactly* what to do it *does not have to be.*

More than anything, stimulation of the clitoris during intercourse depends on the stance or movement of the female, and so is much more under female control than male. This is as true for humans as for any other mammal, and it is not surprising in view of our conclusions concerning the function of intercourse orgasm (Scene 25). This fact makes life extremely difficult for an inexperienced male, as he struggles to learn the techniques of intercourse. If his own behaviour sometimes does and sometimes does not stimulate the woman, a man can only learn how to exploit this most powerful source of sexual stimulation either from a great deal of trial and error or from direct education by a woman. Even then a clitoral technique that works with one woman does not necessarily work with another (Scene 36).

On the night of Scene 27, the girl was looking for a long-term partner (Scene 18), and as part of her selection procedure her body was eager to test her chosen male's competence at intercourse. Because he had at least learned something of courtship, the young man passed her early tests and had a chance of passing her intercourse test. He failed dismally. The more experienced man, however, passed her tests at all levels. The girl probably preferred the older man for many reasons, but one was undoubtedly his superior sexual technique.

When asked why they like men who can stimulate them to orgasm, women naturally reply in terms of the pleasure they gain from orgasm. But the advantages a woman gains from selecting the more competent men are biological as well as sensual. This is despite the apparent disadvantage that, the more influence a man has over her orgasm pattern, the less control she herself has. Yet we have argued (Scenes 22 to 26) that a woman's control in this respect is a major weapon in influencing her reproductive success. Is it possible that this loss of control is more apparent than real?

No man can *force* a woman to climax if her body does not want to. A competent man simply *helps* her to climax when she does want an orgasm. A less competent man forces her to do much more for herself via masturbation and nocturnals. An experienced and competent man is therefore a help, not a threat. But there is another aspect to the female preference for more experienced males.

Basically, a woman uses a man's approach to foreplay and intercourse to gain information about him. A man who is able to arouse a woman and stimulate her to orgasm signals that he does have past experience of other females. This tells her that other women have also found him attractive enough to allow intercourse. The more effectively he stimulates her the more experienced he should be – and hence the greater the number of women who have so far found him to be attractive. Mixing her genes with his, therefore, may produce sons or grandsons who are also attractive to women, hence increasing her reproductive success.

Interestingly, the females of some species of birds are also known to use this yardstick during mate selection. If a female sees one or more other females mating with a particular male, she also is more likely to mate with him. So being seen to be attractive to other females is an attractive male trait in its own right.

Despite the difficulties, virtually all men eventually learn the basics of intercourse, and most learn even the subtleties. But the faster they are able to learn, the fewer opportunities for insemination they will miss in their lifetime, and the more children they are likely to father, with more women, than their slower-learning contemporaries. Studies have shown that boys and girls who experiment most before puberty, particularly if that experimentation involves genital contact, have more sexual partners in their lifetime. Making the most of early opportunities for sexual practice gives a young boy the edge over his contemporaries, and ensures that in generations to come he will have more descendants in the population.

Obviously, failure to practise when younger does not totally condemn a young man to a life without successful intercourse. If a girl or woman finds a man sufficiently desirable as a short- or long-term partner in other respects, she may tolerate inexperience, a lack of consideration, and other sexual shortcomings. Mate selection is, after all, a process of compromise (Scene 18). In Scene 27, the flatmate undertook to educate the young man in the niceties of intercourse. Her aim was to create a partner who, in addition to qualities that she obviously quite liked, would be moderately competent at helping her to climax when it suited her. While educating him, she was also testing his ability to learn sexual techniques. He was becoming 'quite good', suggesting that any sons and grandsons he might give her would also be at least 'quite good'. Importantly, in educating him to help her to climax, she was not educating him to help other women to climax – or, at least, not to the same extent (Scene 36). He would learn something from her, but not as much as it might seem.

In this discussion, we have concentrated on the sexual development of young males as they struggle to pass the tests created by women in the name of mate selection. Although young women have far less to learn about the basics of intercourse than young men, they do need to learn how to recognise and act upon the urges generated by their bodies. For example, a girl has much to learn about how to masturbate and when to do so – or not. She also has to learn how and when to seek a man's cooperation in stimulating orgasm during foreplay, intercourse and post-play. Finally, she has a great deal to learn about the techniques of deception and reassurance within relationships.

Her urges to do all of these things will be orchestrated subconsciously by her body, but the expertise she develops in satisfying these urges will depend on her ability to learn – and to learn quickly. Failure to learn sexual technique quickly is unlikely to make a woman miss opportunities for insemination as clearly as inexperience may do for a man. Nevertheless, it will influence the extent to which she makes the most of the opportunities that present themselves. Not least, the efficiency with which she learns will influence her ability to make the most of sperm warfare.

We discuss this learning process for women in Scene 31. First, we look at the dangers and repercussions of the way in which women test male sexual prowess in the early stages of courtship.

SCENE 28

Rough and Tumble

The four teenagers walked noisily through the summer wood, sheltered at last from the scorching early afternoon sun by the heavily leafed trees. Ahead of them, squirrels skittered off the path and birds flew into the bushes, sounding their alarm calls. Nobody had trodden this path for a while.

One of the two boys had recently passed his driving test and had borrowed his mother's car for the day. They had parked by the side of the road, and for the past fifteen minutes had been walking through the wood on their way to a place known to one of the girls. At first, they had made their way along the narrow path in pairs, the dark-haired couple at the front, the second boy and his girlfriend behind. Each couple walked hand in hand. But when the path widened out, they broke up. The boys had taken off their shirts and, on some pretext, had started to flail each other with them. Then they had run off, chasing each other among the trees. The two girls linked arms and, bending towards each other as they went, whispered comments about the boys' bodies and behaviour, laughing loudly each time one of them said something particularly wicked.

Leaving the two boys to their games, the girls continued round a bend in the path. Ahead of them in a small, sunlit clearing, a narrow, grass-covered wooden bridge carried the path across a river. Had the girls known what was to happen over the next few hours, they might have thought that the dark pine plantation into which the river flowed looked menacing. As it was, they found the place cool and beautiful. The river was shallow, and the pebbles on its bed made the water babble soothingly as it flowed along.

This sound of running water was the perfect antidote to the heat of the day. Even better, just before the river bent under the bridge and into the plantation, the water formed a large shallow pool just over a metre deep.

As the girls approached the pool, they were overtaken by their two companions, now breathless from their exertions. The dark-haired one stopped at the edge, quickly stripped to his swimming shorts and jumped in, immediately at ease in the cold water. The girls were much slower stripping down to their bikinis and he grew impatient, splashing and throwing water in their direction, provoking shrieks of 'No', 'Stop it!' and 'Don't do that!' There was much running backwards and forwards as, each time they approached, he splashed them. Profanities flew amidst the shrieks of shock and laughter. In the end, the two girls jumped in and began to exact wet and cold revenge on their antagonist.

The second boy watched from the river bank and slowly stripped down to his shorts. He had been dreading this moment. He couldn't swim and in truth had a phobia about water. He would have been able to cope on his own, but with all the horseplay going on he was secretly terrified. There was no way he could go into the water with the others. He sat and watched their fun and waited nervously for the inevitable.

He secretly envied the dark-haired boy; envied his muscular bulk, his athleticism – and his ability to swim. He was like a magnet to girls. At parties they would cluster around him, hanging on his every word, laughing loudly at his jokes, each trying to make him notice her. According to rumour he had had sex with almost every attractive girl in their circle, even with the second boy's own current girlfriend. He, on the other hand, was still a virgin. He hadn't even been able to persuade his girlfriend, no stranger to sex, to let him enter her. She didn't mind his fingers inside her and she happily played with him until he ejaculated, but not once in three months had she let him have intercourse.

One of the few from their group who had not so far had sex with the dark-haired boy was the dark-haired girl who was with him today. She was in many ways the most attractive girl in their year at school. Until today, though, she had tended to distance herself socially, and was usually to be seen being driven around by older men. None of these escorts had ever lasted long, and rumour had it that she was still a virgin and something of a tease. She had always refused to go out with anyone her own age, even the dark-haired boy. Today, however, his persistence had been rewarded when at last she had accepted his invitation. As the second boy watched her white bikini go virtually transparent as it became wet, he had to accept that really she was much more attractive than his own girlfriend. As he watched the two girls playing with and making up to the lad in the water, he grew more and more jealous.

The other three knew that he couldn't swim, but under-estimated his fear of water. They urged him to join them, pointing out that it wasn't really very deep. As he continued to decline, they grew irritated and impatient. Even his girlfriend, embarrassed by how badly he was measuring up to the other boy, became annoyed. Although she genuinely enjoyed her boyfriend's company, she was much more attracted to the dark-haired one. A few months ago she had had sex with him, but he had dropped her almost immediately. For a while she had been inconsolable, and she still hadn't really given up hope that one day she would get him back. At this moment she was much more concerned about flirting with him than with her boyfriend. When the other two decided it was time to force her boyfriend into the water, she actually joined them. They scrambled out, grabbed hold of his feet and hands and, despite his obvious fear and genuine pleadings, threw him into the pool. Trying hard not to show his panic, he quickly climbed back out of the water, swore at his companions, then settled back on the river bank.

Eventually, his girlfriend came and sat with him and soon everybody was out, warming up in the sun. After a short, quiet spell he started to throw grass stalks at the other boy. There was another chase, and an uneven wrestling match which he easily lost. When they returned, it was the girls' turn to throw vegetation at the dark-haired boy. He threatened recrimination, went down to the river, cupped his hands to get a handful of water, and threw it at the two girls amidst shrieks of laughter. They retaliated by joining forces and pushing him into the water. Then they jumped in after him and the fray continued.

The other boy sat, watched and envied. Amidst all the splashing and shouting, he heard his girlfriend say she was going to pull down the dark-haired boy's shorts. Soon both of the girls were wrestling with him. As soon as he realised what was happening, he grabbed his shorts, shouting, 'No . . . stop it!' and swearing loudly. The other boy watched, a jealous half-smile on his face, wishing something like this would happen to *him* once in a while.

The dark-haired boy was excited. The idea of losing his shorts to the two girls appealed to him, but he still fought and protested. He yelled, 'No!' a few more times, swore, insisted his shorts would tear, and then when he felt he had protested long enough stopped struggling and let them have their way. It was the other boy's girlfriend who completed the job, waving her trophy excitedly in the air. When he grabbed at his shorts, she threw them to the other girl. The dark-haired boy and the other one's girlfriend collapsed in a heap into the water. Meanwhile, the dark-haired girl climbed out of the pool, picked up her boyfriend's other clothes, and disappeared with them into the woods. The naked boy was so busy playfully wrestling with the other boy's girlfriend and getting aroused that he didn't notice what was happening on the bank – until his own girlfriend reappeared. When he realised what had happened he sank back into the water to hide his erection, while he worked out what to do next.

After listening to the jokes and the accusations of cowardice for not standing up, he turned to his girlfriend, urging her to get his clothes for him. However, despite good-humoured banter and threats, she refused to cooperate. False irritation crept into his voice and, eventually, he announced that he didn't care who saw him 'with a hard-on' and that his girlfriend was going to pay. With that, he swam towards the bank and clambered out of the water. His girlfriend stood for a second, frozen with excitement and indecision. Then, as he stood facing her on the bank, she began to run up the path towards the bridge. The boy gave chase. The pair were only just in view of the others when he caught her, wrestled her to the ground and sat astride her, pinning her arms to the ground. They were both breathless. She was laughing, but he was not. Still visibly excited, he was on the verge of losing control.

He asked where she had hidden his clothes. She said he had to find them. He put his hand on her bikini top and said that if she didn't show him where they were, he would take it off. She pretended to give in, saying she would show him where they were, but as soon as he let her stand up she went to run off again, laughing. He grabbed her, put one hand on the back of her neck, twisted an arm behind her back, and asked her again to show him. She protested that he was hurting her. He wouldn't let go until she showed him his clothes, he said, and began to frog-march her into the wood, heading back to the place where he thought she had taken them. They disappeared into the trees.

The other boy's girlfriend had become aroused by her close contact with the dark-haired boy, memories returning of the occasion they had had sex. Disappointed at his departure, she redirected her arousal towards her boyfriend. She reached out her hand to pull him to his feet, and suggested that they too went for a walk. Once he was standing up, he pulled her cold, wet body against him and kissed her. When she felt him becoming aroused, she suggested again that they move on in case the others came back. They walked a short way until they found a secluded spot on the river bank and lay down hidden from the path by vegetation. When he complained that their wet clothes were uncomfortable and that they should take them off, she agreed. For the first time since they had met they could actually see each other's nakedness, instead of fumbling with each other in the dark of his mother's car. He knew she was on the pill, and assumed that at last they were going to go the whole way. She assumed that, as he had always restrained himself in the past, he would do so again.

At first, their petting followed its usual course. He kissed and stroked her rapidly warming skin. She stroked his back. He fiddled inexpertly with her genitals, she held his penis In her hand. He said he didn't want to come yet, and moved to get on top of her. When she complained, he said he just wanted to rub himself against her. She said she didn't want him to. He urged her to let him, and eventually she relented. After a few seconds on top, he said he wanted to enter her. She said no. He pleaded, she refused. He asked why she had let his dark-haired friend have sex with her but wouldn't let *him*. She hadn't wanted to have sex with the other boy, she lied. Anyway – and she said this with more honesty – he hadn't given her much choice.

It was the last straw. Lying naked in the sun, on top of a naked girl, his penis no more than a thrust away from her vagina, he was more aroused than he had ever thought possible. All afternoon, he had felt inadequate and suspected, correctly, that his girlfriend still preferred his friend to him. Now he learned what her preference was despite the fact that he had actually forced her to have sex. The cocktail of emotions was tipping him over the edge. He wanted sex with her, and he was going to have it. If his friend could force her, so could he. If she could still like his friend despite his forcing her, she would still like him. He moved his body down just the distance necessary to push himself in – and missed.

As soon as the girl realised what he was trying to do, she told him to stop and began to struggle. Briefly he pleaded with her to let him in, but when she still said no he used his weight and strength to pin her to the ground. They struggled as he tried both to hold her down and to find her vagina. He stabbed and stabbed with his penis, but kept missing. Slowly it dawned on the girl that, unlike his friend, he didn't know what he was doing. He had the strength, and the weight, but not the experience. He was trying everywhere but the right place. She carried on biting and scratching and angrily protesting, while subtly changing her position to try to make penetration easier for him. But it was too late. The feelings of frustration and inadequacy merged into panic in the young boy's body. He still wasn't inside her, and he couldn't hold off ejaculation any longer. As the spurts came, the semen running down between her buttocks on to the grass, he felt totally pathetic. He had risked all, and failed.

For a few brief moments, the girl went quiet. Then her emotions took over. The mixture of fear and sexual excitement she had experienced as they struggled, the lack of release through either penetration or orgasm, and the disdain for him that had been growing throughout the afternoon – all erupted in anger. At every step during the last few hours he had dwindled in stature in comparison with his friend. She called him a bastard, told him he was pathetic, taunted him that he couldn't even manage to rape her, and said she was going to report him.

As she pulled on her bikini and moved to stand up, he shrank still further in her eyes. Post-climactic release had removed all trace of aggression and pride, and there were tears in his eyes. He apologised, said it wouldn't happen again, and begged her not to tell anybody. As an afterthought he said he wasn't really trying to enter her, that of course he would have succeeded if he had really meant to, and that it was just a game. He must think she was stupid if he thought she or anybody else would believe that, she told him – he was going to get what he deserved. With that, she threw his wet shorts at him, told him to cover his pathetic prick, and stalked off saying she was going to find the others and tell them what had happened.

Even if she had found them, the others wouldn't have been interested as they had just acted out a drama of their own. The naked dark-haired boy, penis still erect, had frog-marched the girl through the pine trees to the place where she said she had hidden his clothes. She had taken him into the darkness of the plantation. There was a soft bed of sweet-smelling pine needles beneath their feet, and they were hidden from the sky and the path by a regiment of ten-year-old trees. Both were high on the emotions of the past few minutes. The wrestling in the water as they had struggled with his shorts; the chasing, the arm-twisting, their state of undress – had all combined to produce in each of them a high of sexual excitement. There was no question in the boy's mind but that when she had shown him where his clothes were, they were going to have sex. For him, there could be no other outcome. Sexual excitement always aroused in him aggression and an urge to dominate, even hurt. This happened every time he had sex, even without the rough-and-tumble foreplay and teasing he had experienced today. He wanted sex with this girl, but he also had a strong urge to humiliate her in the same way as she had tried to humiliate him.

The girl continued to feed off his frustration, aggression and visible excitement. The rumours about her were true. She was still a virgin. Fingers, but not penises, had been in her vagina. In truth, she had a fear of real penetration, a fear that her vagina would split – the thought had kept her awake many nights. Yet, at the same time, she had a high level of sexual motivation. She masturbated often and enjoyed close contact with the male body. More than anything, she relished her mounting tension as she waited for the moment to deny penetration to a male who had aroused himself with her body. She knew this boy's reputation for roughness and aggression during sex. He had, after all, had sex with nearly all of her friends over the past year. Nevertheless, she was convinced she could handle him.

When she pretended just one time too many that she couldn't remember where she had put his clothes, and then laughed at him, she gave him the excuse he was looking for. Again he twisted her arm behind her back until it hurt, pulled her against him, and asked where his clothes were. Still thinking he was joking, she repeated that she couldn't remember and complained that he was hurting. The next second she was on the ground, face down on the soft bed of pine needles, and he was taking off her bikini, almost standing her on her head in his urgency to pull off her bottom half. For a few moments she was shocked, but still thought it was play. But when he sat astride her buttocks, put both arms up her back so it hurt, and pressed her face into the pine needles so that she could hardly breathe, her excitement rapidly began to fade. He was going too far and hurting her too much.

He told her that if he got his clothes *now*, she might just get her clothes back before the afternoon was over. He grabbed hold of her hair, raised her head off the ground, and asked again. She said OK, but to stop hurting her. When he released an arm, she indicated a fallen branch ahead, about waist high. He dragged her to her feet, once more pulling both of her arms behind her and pinning them painfully against her back. Then he marched her over to the fallen branch, pushed her stomach against it, forced her to bend forward, and asked where. She cried that he was hurting her, pleaded with him to stop, and told him exactly where the clothes were. He saw them, and then told her he would teach her not to mess him about again.

She was hurting where the branch was grazing her stomach and from the way he was pushing her arms up her back with his left hand, but still she didn't realise what he was about to do. So far, everything had been physical, not sexual, and although she was in pain and surprised at how powerless she was to resist his strength, she was still exhilarated rather than afraid. It wasn't until he began to push her legs apart with his feet that fear suddenly gripped her stomach. Almost before she could find the words to ask him what he was doing, he was inside her. She told him she wasn't on the pill and begged him to stop. When she tried to cry out, the sound stuck in her throat. The force with which he was pressing her down over the branch and the pain from her arms, stomach and virginal vagina took her breath away. Each thrust was agony. She began to cry, and again and again begged him to stop. He didn't, but at least he ejaculated quickly.

With ejaculation, his aggression faded. He told her how good it had been and that she was the best yet. He said he was sorry if he had hurt her, lifted her gently off the branch, and tried to pull her towards him to hug her. She resisted. He was a bastard, she sobbed, and he had hurt her. He stroked her head, kissed her tears, and told her that he really hadn't meant to hurt her; that he thought she had wanted it to be like that. For a while, they sat with their backs against a tree, him with his arm round her shoulders. She stopped crying, went very quiet, and began sucking her thumb. She was in pain, both mentally and physically, and at a complete loss as to what to do. Three thoughts kept going through her head: she had been raped; she was no longer a virgin; and her vagina could take a penis after all, even if it hurt.

He spoke intermittently, saying at least three times how good it had been and how much he liked her. Just once, she accused him of raping her. He laughed and said that he supposed it was a bit like a rape. She said it wasn't a bit like rape, it *was* rape. A little later, she said he should have used a condom. He apologised, said he had brought some and that they were in the pocket of the trousers she had hidden. Then he said she shouldn't have got him so excited, and in the next breath tried to reassure her over pregnancy and AIDS. Eventually, they got dressed and made their way back to the river to join their friends.

During the walk back to the car and on the drive home, the two girls and the boy who was driving were very quiet and withdrawn. Only the dark-haired boy chattered as if nothing had happened. Later that evening, with the boys gone, the two girls confided in each other. The driver's girlfriend triggered their conversation by saying that she had been raped. She claimed it was the second time and told her friend about her previous experience with the dark-haired boy. Her friend then described her afternoon.

They compared notes on what the dark-haired boy had done to them. Then they told each other the rumours they had heard about his treatment of some of their friends. They mused over why he forced girls when most of them, given time, would happily have sex with him anyway. Not very seriously, they considered reporting him to the police, to stop him raping others. Perhaps they should both go and report both of the boys for the afternoon's incidents. They talked through the possible scenario at the police station, but decided that the experience might be worse than the actual rape. Besides which, it would mean their parents would find out and they would probably never be allowed out again.

The driver's girlfriend had no more to do with him, and within days began going out with another boy who had his own car. Whenever she met him, she snubbed him or called him a rapist.

She never missed an opportunity to tell her friends how pathetic he was and how she couldn't understand why she had gone out with him in the first place. Once away at college, he rarely came home, anxious to avoid the people who thought so little of him.

The day after the other girl lost her virginity, she went on the pill and, after turning the dark-haired boy down twice, eventually agreed to go out with him again. They carried on seeing each other for the rest of the summer, the longest he had ever stayed with anybody. She gained the appropriate kudos and aroused the jealousy of her friends for having landed the boy they all wanted. Despite occasional infidelities while at college, they continued to see each other on and off for the next three years. Eventually they began to live together.

Their first intercourse in the pine wood set the style for their sexual relationship. In the years that followed, they re-enacted that and similar scenes with mutual consent. For them, intercourse was almost always aggressive and often painful. They enjoyed devising outrageous scenarios to generate fear and humiliation as a prelude to it.

Four years after beginning to live together, they had their first child.

~

Rough-and-tumble sex play is a common element in the courtship of humans and of many other animals, as they make decisions about whether to have intercourse or not. Such behaviour has many facets, most of which are illustrated in the scene – and all of them involve an interplay between mate selection by females and the display of quality by males. The females set tests of physical strength and sexual competence (Scene 27), which the males then either pass or fail. In their pursuit of reproductive success, the judicious use of rough and tumble can have important benefits for a woman – and a satisfactory performance can be equally beneficial to a man.

On the vast majority of occasions, such rough-and-tumble games unfold without any harm being suffered by either the woman or the man. In fact, they both gain. The woman gets the information she wants and the man, if he behaves satisfactorily, may be allowed to have intercourse. Occasionally, however, such games can be dangerous. It is a short step from rough-and-tumble intercourse, across the boundary of mutual consent, to rape. This is date rape – when intercourse is forced by a man on a woman who has at least found him attractive enough to 'date', and often attractive enough to kiss and 'pet'. It is not the random, predatory rape by a man who is usually a complete stranger to his victim (Scene 33).

In principle, it should be easy to draw an unambiguous line between rough-and-tumble intercourse and date rape. If a woman says no, but the man forces intercourse anyway, then the intercourse was a rape. However, as all legal systems round the world acknowledge, the situation is not that simple. One of the problems is that, in many aspects of life, people often say no when what they really mean is 'See if you can persuade me.'

In Scene 28, we saw five incidents in which people said no. On two of these occasions they appeared to mean it, and on three they didn't. First, the girls told the dark-haired boy not to splash them, but they didn't mean it and within a few minutes were enjoying their retaliation. Secondly, the other boy had a genuine fear of the water, said no when invited to join the others, and pleaded not to be thrown in. The other three underestimated his fear, ignored his pleas, and threw him in. Thirdly, the dark-haired boy at first did not want to lose his shorts, said no, and fought to keep them. He soon changed his mind and decided that he would like to lose them. Nevertheless, he continued to say no and to struggle. Fourthly, the other boy's girlfriend resisted her boyfriend's attempts at intercourse. She said no and struggled. But when she realised he wasn't capable of forcing intercourse she changed her mind at the very last minute and tried to help him penetrate. Yet she continued to struggle and say no. Finally, the dark-haired girl had a genuine fear of penetration. Although she had enjoyed the rough and tumble, she said no at the first indication that the boy intended to force intercourse, and struggled and pleaded thereafter as best she could. Underestimating her fear and not believing that she really meant no, he ignored her pleas and inseminated her.

At first sight, the second and fifth of these events fit into one category, and the remainder into another. Unfortunately, the situation is not that simple. As far as date rape is concerned, there is a further complication – the reaction of girls who have been date-raped in the weeks that follow their experience.

In a study of American students published in 1982, it was found that girls who were exposed to an attempt at date rape were three times *more* likely to resume their relationship with the man concerned if his attempt *succeeded* than if it failed. Presumably, the fact that all of these women had claimed the man had attempted rape means that at the time they said no and meant no. Yet, if the man succeeded in forcing intercourse, nearly half (40 per cent) later resumed their relationship with him, just like the dark-haired girl in the scene. If he failed, then nearly nine out of every ten of the women (87 per cent) refused to have any more to do with him, just like the girlfriend of the less popular boy in the scene.

This reaction of women makes it even more difficult to draw the line between date rape and rough-and-tumble intercourse. In the discussion that follows, we shall make no further attempt to distinguish between the two – that is a job for the legal profession, not a biologist. The phenomenon under discussion is rough-and-tumble intercourse, and the arguments that follow are concerned with the influence this behaviour may have on male and female reproductive success.

In Scene 27, we discussed how much men and women have to learn in order to get the most out of their earliest sexual opportunities. We also discussed the link between learning and mate selection, particularly the way that a female needs to set tests of competence for males as part of her process of selecting.

This testing is an important process – a woman uses it to identify which men will give her sexually competent sons and grandsons. But when such testing involves rough-and-tumble sex play, as in Scene 28, it can become a dangerous business. Nevertheless, even in the scene, both girls survived relatively unscathed and both came to a decision about their respective partners. To judge from each girl's subsequent behaviour, one of the boys passed that afternoon's tests, the other failed. So what *were* these tests, and what insight do they give us into the function of rough-and-tumble sex play in humans and other animals?

Before we can answer this question, we have to be clear over precisely how a man and woman's interests differ when they begin such behaviour. Many of the important factors are touched on elsewhere in this book, but in the discussion that follows they become crucial – and this is because here we are primarily concerned, as in the scene, with a couple's first intercourse.

Once a man and woman have established a long-term relationship, the costs and benefits that they each experience from intercourse become similar (Scene 16), even though the function may differ (Scene 2). But with first-time intercourse the situation is very different. Apart from the chance of contracting disease, which is a risk they share, their potential costs and benefits are not at all the same – especially if there is a strong possibility that their first intercourse together may also be their last. A man, like all male animals, has much less to lose and much more to gain from a one-off intercourse than a woman.

From the viewpoint of reproductive success, and disregarding here any social pressures to which the man may be exposed, the siring of a child with a woman other than a long-term partner need not be expensive for him (Scene 13). Having inseminated her, he can if he chooses attempt to avoid all further contact with both her and the resulting offspring. A single intercourse need have no greater cost than that involved in fending off future claims for help and support if she does have a child. Against what could be, therefore, a relatively minor cost he can set the chance of producing a child. If he does not take each opportunity to inseminate a new woman as it arises, he may never get another chance. In which case, whoever fathers her next child, it will not be him. A man's reproductive success depends to a significant extent on his ability to make the most of one-off opportunities.

The situation is very different for a woman. For her, conception is a major event. It may commit her at least to months of pregnancy and usually to years of dedicated effort. The man concerned might desert her (Scenes 8 to 11, and 16). Moreover, conceiving via a man later proved to be genetically inferior could result in her raising a less successful child than if she had waited for a more suitable man (Scene 18). Together, these two dangers mean that an incautious, one-off intercourse can considerably reduce a woman's reproductive success. Her priorities need to be when (Scene 16) and with whom (Scene 18) she has one-off sex, rather than how often. Caution and selectivity are of maximum importance.

Of course, women are not always cautious. The primary situation in which a woman might abandon caution, though not usually selectivity, is when she has a one-off opportunity to collect sperm from a particularly desirable man. Usually, this is a man who she has already judged from a distance would make a good genetic father for her next child (Scene 18). In this book, we have seen several women behaving in this way (Scenes 6, 17, 19, 21 and 26). Most often, as in those scenes, such behaviour is within the context of infidelity. Once a woman has a long-term partner, the costs of one-off intercourse are reduced as long as her infidelity remains undetected (Scenes 9 to 11). Her long-term relationship provides a springboard from which to exploit the genetic benefits of one-off sex with selected men without risking too much. She does not have this freedom, however, if she does not have a partner.

The result of these pressures is that women, like all female birds and mammals, are genetically programmed to be cautious and selective. In past generations, women who were not so were less successful reproductively than those who were. All women alive today are the genetic descendants of the more cautious of female ancestors, not their more reckless contemporaries. Men, on the other hand, are genetically programmed to be urgent and single-minded about one-off sex. In past generations, men who were not urgent and persuasive were less successful repro- ductively than those who were. All men alive today are the genetic descendants of the more urgent of male ancestors, not their more complacent contemporaries.

We can see, then, that there is a big difference between men and women in the potential costs and benefits of a one-off inter- course. This difference automatically means that the two sexes approach any potentially sexual situation with a conflict of interests.

A man can only satisfy his urgency if he can convince the woman concerned that she actually wants to be inseminated by him, now, rather than by him or somebody else, later. His only alternative is to try to force insemination whether she wants it or not. At first glance, it might seem that both of the boys in Scene 28 simply opted for the second course of action. Again, the situation is not that simple. The complicating factor is that male persistence in the face of female resistance can be a normal, mutually acceptable facet of courtship and foreplay. So, too, can aggression and a level of physical trauma. We are back to the way that women set men tests as an aid to mate selection – which brings us to the function of rough-and-tumble sex play.

This is an emotive issue – so emotive that it is probably best broached by first considering the courtship of animals other than humans. For example, if we watch the courtship of dogs, we often see a clumsy male persisting and persisting with his advances despite rejection after rejection by the female. If we watch domestic cats we see females clawing, scratching and spitting at prospective suitors. If we watch mink, we see a male drawing blood as he tries to subdue the female's spirited resistance.

Witnessing such behaviour, it is difficult not to feel sorry for the females. At best they are being pestered, and at worst they are being physically damaged by males who refuse to take no for an answer. Yet, despite their resistance, female cats and dogs do eventually allow one of their persistent and aggressive suitors to mate. As for female mink, if they *do not* experience physical trauma at the male's hands, they *do not* ovulate. Their bodies are on hold, waiting for the right male to inseminate them before they produce an egg (Scene 15). In all of these animals, female resistance is actually a test of male competence. Rough-and-tumble sex play in humans is similar.

On average, men who are physically able to overcome the final defences of a female and achieve insemination leave more offspring than those who are not. So women whose sons and grandsons also have this ability will enjoy greater reproductive success. One of the criteria (Scene 18) that a woman can add to her list when selecting a mate, therefore, is his ability to overcome her physical resistance – but how does she test such an ability?

Initially, she can simply watch him in competition with other males. The young men in Scene 28 spent much of their time chasing and wrestling with each other, doing their best to display their strengths and hide their weaknesses. But finally, the only real test a woman can set is whether a man can negotiate and overcome her own defences. To test this, she has to resist first verbally, then physically. The stronger and more realistic her resistance, the better the test.

This is, of course, a dangerous game. Resist too little, and the test is meaningless. Resist too much, and the male may inadvertently cause real rather than superficial damage. The fact that courtship rough-and-tumble aggression rarely results in serious damage in cats, mink or even humans shows the accuracy with which this feature of sexual behaviour has been moulded by natural selection. Even in mink, the level of trauma that stimulates a female to ovulate is set at a level just high enough to test the male's ability to overcome her defences, but not high enough for her to suffer long-term damage.

In species such as humans which form long-term relationships, rough-and-tumble sexual behaviour is most important during the early stages of courtship. Once a woman has tested a man's ability to force himself on her, she need not do it often thereafter. But as in all such tests of male health and ability (Scene 20), even within a relationship a woman gains from occasional reappraisal of her partner.

Of course, in this as in most features of human sexuality, people vary (Scenes 35 to 36). For some, rough-and-tumble sex play is indeed a minor and infrequent element in their relationship. For others it is an essential element even up to the level of sadomasochism, if they are to accept the other person as a suitable partner. The dark-haired couple in our scene clearly leaned in this direction. Their intercourse on that first afternoon was rough, painful and humiliating for the girl, yet through the experience she recognised the boy as a compatible partner. In the years that followed, their sex life was to continue in the same vein – even once he was her long-term partner, she tested his ability to force her on many occasions during routine sex.

We can now understand the decisions made by the two girls in the scene in the days and weeks that followed the events by the river. The boy who forced the dark-haired girl into intercourse passed her tests. She was already attracted to him, as were most other women, and her body perceived that his qualities made him seem a good candidate for giving her reproductively successful sons and grandsons. These qualities included physical power and sexual competence, qualities that her own characteristics made her well suited to test.

In contrast, the other boy lost stature in his girlfriend's eyes at all levels as the afternoon wore on. In part, he suffered by comparison. Mainly, however, he failed in an absolute sense to live up to his girlfriend's criteria for acceptability. There were probably two ways he *could* have passed her tests, but he failed in both. The basic cause of his double failure was inexperience. First, with more experience, he would probably have opted for restraint and might eventually have gained long-term benefits (as, for example, did the experienced man in Scene 20 who, when faced with a similar situation, opted for patience rather than force and six weeks later reaped the reproductive benefits (Scene 26)). Instead, however, he opted for force. Even then, with greater experience at intercourse he might still have cemented a longer-term relationship and hence won many future opportunities for insemination.

SCENE 29

How to Con

Outside the car, it was very dark and very cold. Inside, it was warm and getting warmer. After pulling off the road among the trees, he had left the engine running and the heater on while they climbed into the back seat. Now everything was warming up nicely. The girl's breasts were exposed, her knickers were around her knees and his hand was between her legs in the warmest place in the car. She was struggling with the front of his trousers, and he was very excited. As he kissed and nibbled the side of her neck, his ear rubbed against the car window, cold and wet with condensation.

Since acquiring his own car, six months ago, this was the third time he had been in this position. But on the first two occasions, each with a different girl, he had missed his chance.

The first time, he had naïvely assumed that any girl interested in sex would be on the pill. She wasn't – and she wouldn't let him have sex without a condom, which he didn't have anyway. He had pleaded with her to let him in, promising he would withdraw before he came. She said she had been promised that once before, would never trust a boy again, then immediately lost interest and asked him to drive her home.

The second time, he had simply been taken by surprise. Offering a lift to a girl he scarcely knew, he had been amazed to find that before they were even half-way home he was being seduced into finding a quiet spot for them to enjoy themselves. They had been on the verge of intercourse when she suddenly stopped and asked him to use a condom. When he said he didn't have one, she pushed him away. He offered to drive off and buy some, but she too had lost interest and had wanted to go home.

With both girls, he had only the one opportunity for intercourse. Neither had given him a second chance. He had vowed never to miss another opportunity, and ever since his second experience had carried a condom with him. Now, two months on, with the foil around the condom looking faded and worn, here at last was his chance to reap the benefits of his hard-earned lesson. If he failed this time, it wouldn't be because he wasn't prepared – or so he thought.

The moment came, and without even being asked he fumbled for the condom in his pocket. As he tore off the foil, the girl took off her knickers and got herself into as comfortable a position as she could. He pulled the condom out of the foil, perched it on the tip of his penis, and tried to roll it on. But it wouldn't go. The condom just wouldn't roll. He held it up in the blackness to try to work out which way round it should go, but could see nothing. He turned it over and tried again. He began to panic as he felt movement deep in his penis. The girl asked if he was having trouble. It was fine now, he lied, and with the unrolled condom still balanced precariously on the tip of his penis he moved into position and entered her without help.

He knew the condom had come off as soon as he started thrusting, but he had been waiting for this moment too long to stop now. With supreme will-power he fought off the ejaculation that was so near as he entered and revelled in the excitement of an unprotected intercourse. Not until he went to withdraw, several minutes after ejaculation, did he tell the girl the bad news, feigning surprise as he pretended to search for the condom end with his fingers. The girl swore and, in panic, tried to find the condom inside her. In the end, with longer fingers and a better angle, it was he who found and removed the still unrolled rubber. He apologised profusely, admitted the possibility that he might not have put it on properly, but suggested that perhaps they had just been too vigorous in their love-making and she had pulled it off him.

As he drove her home, he tried to argue that there wasn't really any danger. The condom would still have stopped him from shooting the sperm into her womb and it did have a chemical on it that killed sperm. Even if a condom came off, he argued, it still worked like a cap. Being relatively naïve herself, the girl believed him. In the days that followed, he bought a supply of condoms and practised. At great financial expense, he practised until he could don a condom in any light, in any position, and with either hand.

He never accidentally lost a condom during intercourse again, but did so deliberately on five further occasions, each time with a different girl and each time after they had once previously had intercourse with the condom properly in place. Each time, frustrated by the reduced sensations of wearing the rubber, he rolled the condom down such a short way that it came off almost immediately he began to thrust.

None of these deliberate deceptions produced a baby. Four of the girls did not ovulate after their 'accident'. The fifth did ovulate and her egg was fertilised. However, the 'accident' with the condom and the prospect of pregnancy and motherhood just before the most important exams of her life stressed her so much that, when the fertilised egg reached her womb, it passed straight through without implantation. When her period began, she went out to celebrate.

Nevertheless, the young man's misuse of condoms did make him a father, but through accident, not design. The mother of his child was the girl with whom he had accidentally first discovered the condom's potential for gaining unprotected intercourse. Naïvely reassured by his story of the condom working like a cap, she had patiently waited for seven weeks for a period that never came.

∼

In Scenes 27 and 28 we have explored the way that young men and women have to learn various techniques in order to make the most of their sexual opportunities. In the modern world, they also have to learn how to use contraception. The young man in the current scene missed two opportunities for intercourse through his general lack of experience with girls. He nearly missed a third through his lack of experience with condoms.

When we discussed family planning in Scenes 16 and 17, we concluded that modern contraceptives may have little impact on the *total* number of children a woman has in her lifetime. Nevertheless, they do supplement her natural methods of family planning, thereby giving her even more control over when and with whom she conceives. For a woman today, modern contraceptives are an important aid in her pursuit of reproductive success. Not least, they are a useful weapon in her manipulation of sperm warfare. In this section, we discuss the ways in which a man may also use modern methods of contraception to enhance his reproductive success.

The idea of hindering or killing sperm as they leave the penis is not new. Over two thousand years ago, Pliny suggested rubbing sticky cedar gum over the penis before intercourse. The sheath has been known since Roman times and was in use in many parts of Europe by 1700. Fallopio designed the first medicated linen sheath in the 1500s, but the item took its name from the personal physician to King Charles II, the Earl of Condom, who recommended its use to the king as an aid to prevent the contraction of syphilis. By the 1890s, all of the barrier methods of contraception in use today were openly on sale in the United Kingdom. However, their use is unlikely to have been widespread until well into the twentieth century. In the 1980s, about 50 per cent of couples in countries such as Britain were relying on the man for their contraception. Only 30 per cent were using condoms; the remaining 20 per cent were simply using withdrawal.

The perhaps surprisingly high incidence of withdrawal as a means of contraception during routine sex is in part a reflection of how men feel about using condoms. As most women can testify, if a man is expected to wear a condom during intercourse, he becomes much more cavalier in his attitude towards contraception. Of course, when asked why they don't like using condoms, men say that they simply don't enjoy sex so much with them on – the 'Wellington boot' syndrome. Women, on the other hand, tend to be much more favourably disposed towards them. This difference between the sexes in attitude towards condoms reflects a very important consequence of their use – by preventing sperm from entering the vagina, condoms negate a man's reproductive benefit from intercourse much more than they negate a woman's.

This difference is less marked during routine sex than during casual sex. When a man and a woman are in a long-term relationship, the timing, spacing and number of children that are best for one partner are usually also best for the other. Consequently, an untimely conception would be equally disadvantageous for both of them (Scene 16). Since the use of condoms is one way that both partners can avoid conception, we might expect both the man and the woman to be equally appreciative of their use. Even in a long-term relationship, however, there is some difference between a man's and a woman's liking for condoms. Why?

The main reason is that, as we have already discussed (Scene 2), conception is not the main function of routine sex. Such sex is the means by which a woman hides her fertile phase (Scene 2) while her partner tries to keep her topped up with his sperm as a defence against sperm warfare (Scenes 2, 4 and 6). Condoms do not spoil a woman's subconscious rationale for having routine sex, but they do spoil a man's. A woman can still hide her fertile phase via routine sex whether a condom is used or not. Obviously, however, if the sperm a man ejaculates do not take up residence in the female's tract, they provide him with no defence against sperm warfare.

It is not surprising, then, that even during routine sex men are consciously and subconsciously less enthusiastic about the use of condoms than women (Scene 17). Not least, men are much more likely to advocate taking the occasional risk.

To understand why this difference in attitude is even more marked during 'casual' sex, let us first consider the pressures on a woman during casual sex and how these are affected by the use of a condom. As we have already discussed (Scene 28), a woman is normally more cautious and selective over casual sex than a man. All the same, having decided on a suitable time, place and partner, she can still gain a number of advantages from casual intercourse – as long as she doesn't conceive (see Scenes 18, 20, 27 and 28). For example, if she is without a long-term partner, intercourse can help her gain a man's attention in her search for one. In addition, she can use the intercourse to gauge a man's sexual competence, potency, and to some extent health and fertility. Casual intercourse can thus be an avenue to protection and financial or other help from a man who she judges might be a suitable long-term partner. If she already has a long-term partner, casual sex with somebody else can provide a 'reserve' – a man to move on to if her current relationship breaks up (Scenes 16 and 19). None of these benefits from casual sex require her to conceive. In fact, she retains more options if she does not conceive. Only when she is specifically after a particular man's genes (Scenes 6 and 26), or when she is trying to manoeuvre him into a long-term partnership by becoming pregnant (Scene 18), does she benefit from conceiving as a result of casual intercourse. Except on these occasions, therefore, a woman gains from the use of a condom during casual sex. Condoms even help to reduce one of the potential costs of casual sex, the risk of infection.

Of course, a man also benefits in this last respect. Otherwise, though, the pressures on a man are quite different from those on a woman. We discussed these pressures in detail in Scene 28 in explaining why men are much more urgent, single-minded and cavalier over casual sex than women. Briefly, intercourse with as many women as possible is one of the main ways whereby a man can enhance his reproductive success. Each child produced in this way is a bonus to be added to that mainstay of his success – the children of a long-term relationship. A man suffers very few costs from casual sex that might erode this potential bonus. Harsh though it might seem, all each conception needs to cost him is a few minutes of his time, an ejaculate, and a slight risk of contracting a disease. If a longer-term relationship with the mother is in his reproductive interests, he may decide to offer paternal care. But if it is not, he may simply leave her to raise the child (or not – Scene 16) as best she can, while he pursues other sexual opportunities and seeks a more promising long-term relationship.

The major problem a man encounters in trying to bolster his reproductive success via casual sex is the difficulty of finding enough women to cooperate with him. Few, if any, men get as many opportunities for casual sex as they are programmed to seek – and hence would like. When such opportunities do arise, they are programmed to let as few pass by without insemination as possible (Scene 28). Using a condom to remove all chance of conception thus negates a man's fundamental reason for pursuing casual sex. Subconsciously, his body realises the *futility* of casual sex with no chance of conception – just as, again subconsciously, a woman's body realises the *power* of casual sex with no chance of conception. Yet, despite the apparent futility of their actions, men are sometimes prepared to wear a condom on such occasions. Why?

One possibility is that a man's body is actually fooled by this relatively recent invention into doing something that is against its reproductive interests. It is programmed to assume that when he ejaculates inside a woman the sperm will do their job. Perhaps, despite the evidence from his conscious brain, his body refuses to accept that a condom inevitably negates this assumption.

There is some evidence for this. Men still make all the same topping-up (Scene 4) and warfare (Scene 6) adjustments to the number of sperm they ejaculate during intercourse whether they are wearing a condom or not. They probably ejaculate about 10 per cent fewer sperm if they are wearing a condom, but the adjustments are the same. This suggests that a man's body still 'thinks' his ejaculated sperm may have a job to do – even if it is merely 'wishful thinking' that on this occasion the condom may have accidentally come off or burst.

Even if this were an error in the programming of the male body, it would not be so surprising. Relatively few generations of men have been exposed to the use of condoms. Consequently, the generation game of natural selection (Scene 1) has had very little time to re-program the male body appropriately. But even if we found that men actually *reduced* their reproductive success through their use of condoms, we should expect things to change over the generations to come. Eventually, the human population should become dominated by the descendants of men who used condoms to *enhance*, not reduce, their reproductive success.

At first, it seems counter-intuitive that condoms could ever enhance a man's reproductive success. Nevertheless, there are at least three ways in which they might.

One way would be if a man could use condoms to trade opportunities with a woman. He could offer to prevent conception when they first have sex in exchange for the opportunity of unprotected sex in the future (similar to the situation portrayed in Scene 20). He could then use the protected occasions to convince the woman that he really is a suitable mate. As a result, she might consider risking unprotected sex with him at some time in the future.

Another way would be if the greater protection against infection that condom use affords more than made up for any missed opportunities to inseminate. Since the advent of AIDS, the power of condoms to reduce the risk of disease has become widely known. It is possible that a man who uses a condom strategically throughout his sexual life will on average stay healthier and thus enjoy a greater reproductive success than a man who does not.

The third way would be much more devious – to (mis)use the condom to con women into being inseminated in a way that allows some chance of fertilisation. If a hundred couples used condoms properly for a whole year, no more than three of the women should become pregnant. Yet, in practice up to twenty to thirty conceive. This is nearly half the number (seventy-five) who conceive if they use no modern form of contraception at all. The most likely explanation for this relatively high failure rate is that condoms are not used properly during routine sex. Whether the failures are true accidents or whether the men concerned, like the young man in Scene 29, deliberately use condoms carelessly is not known.

Whatever the explanation for failures during routine sex, the failure rate will almost certainly be higher during casual sex. There is little doubt that many men use condoms during casual sex in all three of these ways. On several occasions, the young man in the scene misused them to give himself a chance of fertilisation during intercourse. On top of that, he used condoms to increase his number of casual encounters. Through not having had a condom to hand, he had missed two opportunities at intercourse, and never had another chance to inseminate either of the two girls concerned. With a condom in his pocket, though, he was able to take full advantage of opportunities with six other girls. On one of these occasions, he produced a child which the girl was probably going to be left to raise as best she could. Quite possibly, he would not have had this child or any of his other opportunities had he not offered to wear a condom.

Maybe the young man in the scene is not so unusual. There is an intriguing possibility that on average men are already using condoms to enhance rather than reduce their reproductive success. If so, their bodies' strategy is as follows. First, try to have unprotected intercourse whenever possible. Secondly, offer to use condoms, as a strategy to increase the number of opportunities for intercourse. Thirdly, while wearing a condom, continue to make all of the necessary topping-up (Scene 4) and warfare (Scene 6) adjustments just in case the condom bursts or comes off. Finally, on occasion, deliberately misuse condoms to achieve 'sneaky' inseminations. Perhaps, through this strategy, learned and instinctive dispositions have already combined to allow the men who are prepared to use condoms to achieve a greater reproductive success than men who are not.

Similar arguments can be advanced to explain withdrawal as a male strategy. Of course, as far as casual sex is concerned the offer of withdrawal gives a man far fewer opportunities for insemination than the offer of a condom. As in Scene 29, a woman is far less likely to accept the offer of the former than she is the latter. There are two main reasons. First, withdrawal offers far less protection against getting pregnant than a condom. Secondly, it provides less protection against infection. As far as fertilisation is concerned, once a man has succeeded in penetrating through offering to withdraw, it is far easier for him to inseminate and fertilise than if he uses a condom. This is illustrated by the fact that withdrawal as a means of contraception has an even higher failure rate than the condom. If our hundred couples used withdrawal for a year instead of condoms, only seven would conceive if they used the method properly. In practice, however, up to forty women would conceive. In part, these failures happen because the few sperm that leak out of the penis before the man withdraws are particularly fertile. Mainly, they happen because men often fail to withdraw as promised.

Offering to withdraw rather than to use a condom provides a man with fewer opportunities for intercourse, but a better chance of fertilisation when those opportunities do arise. Quite possibly, the relative success of the two ploys depends primarily on how experienced is the woman. As the young man in the scene discovered, once a woman has realised that she can be deceived by the promise of withdrawal, she is far less gullible. In fact, most women are rarely deceived more than once, either by the false promise of withdrawal or by a carelessly positioned condom. When she has been conned once, she is likely thereafter to keep a close eye on the whole operation.

Although, naturally, humans are the only animals to use condoms, they are not the only animals to use withdrawal. Many apes and monkeys are known to penetrate and thrust without ejaculation. To what extent this behaviour is simply the male using his penis to remove material from inside the female's vagina

(Scene 21), and to what extent it indicates a tacit agreement between the male and the female not to inseminate – as in humans – is not known. Whenever it is the latter, however, we can be certain that from time to time female apes and monkeys are also conned by males, just like their human counterparts.

10

One Way or Another

SCENE 30

Best of Both Worlds

The young teacher looked up from his marking as his two daughters, aged six and seven, skipped over to kiss him goodnight, herded from behind by his partner like two frisky lambs. One of the girls wished him goodnight in her mother's native language, one in his. Most of the time the household spoke his language but in moments of tenderness or anger any one of them might slip into the mother's tongue. When she returned from taking them to bed, he abandoned his marking, switched on the television and sat next to her on the settee. She put her legs over his knees and they sat, eyes on the screen, idly stroking each other's hands and legs. He looked sideways at her, stomach churning with a fear that had been with him all day. How could he tell her that their whole world might be on the verge of collapse?

'I am Preggnent and its Yours,' the scribbled note had said, thrust into his hand during the morning break. 'My dad says you must giv us sum money or well tell on you.' The girl was fifteen and only slightly more illiterate than most of her contemporaries at the school. Precocious and from a tough family background she nevertheless had the looks and bearing that would fire most men's fantasies – until she spoke. She had asked him for a lift one hot summer's night after school. Foreign languages were his main subject but he doubled as a Physical Education teacher. It was hot that night and he hadn't bothered to change out of his shorts before getting in his car to set off home.

He should have said no when she asked for a lift. He should have told her to stop when she began to stroke his thigh as he drove. And he should have told her to get out when she took off her knickers. But he didn't. Instead he drove to waste ground by a disused warehouse, got into the back seat, and allowed her to pull down his shorts and climb all over him. Looking back, he remembered how expertly she had handled his clothes and his genitals and doubted that she was a stranger to sex. He also doubted that he was the father – assuming she really was pregnant, of course. Maybe she had set up other men, other teachers, then tried to blackmail them also. Maybe the threat was hollow and nothing would happen if he ignored it. But maybe she and her father were serious and he would have to either pay up or own up.

The fear gnawing at his stomach grew stronger. He wasn't yet thirty and his finances were in a terrible state – still not recovered from his years at college, his year abroad, and his year of teacher training. He was still sending money abroad to support the child he had fathered during his exchange year, just before meeting his present partner. Although she had managed to get a part-time job once their youngest daughter had started school, they were still in debt. He wouldn't be able to pay a blackmail demand. His only option, if the girl was serious, was to deny ever having had sex with her and hope that nobody had seen them that evening.

In truth, it wasn't the note that really worried him. If it was merely a case of his word against hers he would have little to worry about, financially or legally. Nor was he unduly worried about how his partner would react. Theirs was a free-and-easy relationship and she had the sexually liberated attitude much more typical of her culture than his. She knew of his child in her home country and, despite their financial situation, had never suggested he should not provide its mother with some support. Though she didn't know the whole story, she knew he had not been totally faithful to her. A combination of bad luck and imprudence had led her to find out about the first of his three infidelities. But she had forgiven him and within a month they were back to normality. He suspected the same would happen again over this schoolgirl, especially if he denied the whole thing completely.

No, his stomach was not churning from fear that his partner would discover his past indiscretions with women and girls. It was churning from fear that she would discover his philanderings with men and boys. His panic was that in defending himself against the girl's note, he might provoke people who knew the other half of his very active sex life into talking. Even apart from the legal dangers, which were serious, he doubted that his partner would stand by him if she knew the whole story, despite her sexual awareness and understanding.

His homosexual activity went back as far as he could remember. He was about six years old the first time his uncle, his regular baby-sitter, climbed naked into his bed to cuddle him. The games they played were 'their secret' which his mother 'wouldn't understand' and which would make her 'very angry'. In any case, he enjoyed them. He liked the smell and feel of his uncle's body, the way their genitals reacted to each other's touch, and the way he could help his uncle discharge. He was ten when he first experienced penetration and twelve when he first penetrated his uncle.

His first homosexual experience with someone his own age was with his cousin. They were both eleven when, playing in his bedroom one day while his parents were out shopping, he had persuaded his playmate that they should take off their clothes and wrestle. Within minutes they both had erections and within the hour he was educating his cousin in some of the techniques he had learned from his uncle. By the time they were thirteen and both able to ejaculate they were meeting at least once a week for mutual masturbation and anal sex.

He was still thirteen when he first had sex with a girl. A young neighbour, the same age as himself, whom he had played with throughout his childhood, had succumbed to his 'I'll show you what I can do if you'll show me what you can do' routine. And he had a lot to show her. The years of practice with his uncle and cousin had given him a knowledge of the body and a sureness and confidence of touch that most males do not acquire until twice his age. The young girl came back, week after week, to have him stroke and stimulate her, each time allowing him to go just that little bit further. In the end, he was touching and caressing her naked body, anus and genitals with his hands, mouth and penis at such length and so exquisitely that her young body was having its first tingling experiences of orgasm. Each week her climax was just that little bit stronger, and each time they had the opportunity she wanted more of the same. After three months he convinced her that she couldn't have a baby from anal intercourse and that she would enjoy it, and after several experiments, she did. Before the year was out, she had asked him for vaginal intercourse as well and by the time he was fourteen his sexual education was complete.

At college, four years later, he joined the gay scene, spending many of his evenings in gay bars and clubs. For a year, he lived with another man in a more or less monogamous relationship, though both of them were occasionally unfaithful, mainly with other men but from time to time with women also. There was never any shortage of women around him, many of whom told him how much better a lover he was than his straight contemporaries.

For his exchange year, he went abroad to study the foreign language he intended to teach. The person in the flat next to his happened to be a woman. They were lovers within days, even before he had found the local gay community. And within three months she had conceived – but not before he had been unfaithful to her with the woman who was to become his long-term partner. Before the end of the year, she too had conceived. Even while living with one woman and being unfaithful with another, he still managed from time to time to slip away for casual homosexual sex without either woman knowing. On balance, though, most of his sexual activity while abroad was with women.

When he returned to college to begin his final year of study, he renewed his activities in the gay community. He kept in touch with both mothers of his children by letter, sending what little money he could. After graduation, he moved to a new city to train as a teacher. No sooner had he arrived than the woman who was now his partner turned up on his doorstep with their five-month-old daughter. She moved in with him and they had lived together ever since. Occasionally, he still picked up a casual lover at a gay bar, but his partner had never found out and as far as he knew she had never suspected his bisexuality.

Six years ago, he had landed his job as a teacher in a rough and depressed town with a high level of unemployment. He vowed to himself that he would never do anything sexually that might prejudice his career or make it more difficult for them to get out of debt. And for three years he had managed to restrain himself. But then, in successive years, he had succumbed to the charms of two student teachers. The first, a young girl, his partner had detected. The second, a young man, she did not. Both affairs lasted only a month. Then, a year ago, he had responded to the tentative advances of another male member of staff and had begun a long-term homosexual relationship, which was still continuing.

Under cover of playing squash, the two men had sex at least weekly in the other man's flat. It was uncanny how similar his new lover was in his moods, behaviour and responses to the girl he had impregnated within a few months of arriving abroad. In addition to sex with each other, the men would sometimes visit gay bars to pick up other men or explore the streets and toilets for young male prostitutes. For these diversions, they drove to the anonymity of the nearest large city, about thirty minutes away.

The problem that was now haunting him began six months ago. With his male lover, he had visited the city to look for child prostitutes. They had just picked up two boys when they chanced across two thirteen-year-olds from their school who were soliciting. It was difficult to know who was the most unnerved by the encounter. His weekend was tense, as he worried about what might happen. He relaxed a little when, back at school on the Monday, the two boys approached him and said that they wouldn't tell if the teachers didn't tell.

Such an alliance might have been stable if the other teacher hadn't then begun inviting the boys to his flat and paying them for sexual favours. One evening, when he called round on his lover, the two boys were still there. The temptation of such a foursome was too great for him, and he allowed himself to be recruited into their games of sexual musical chairs. From then on, the four had met for group sex more or less weekly. Sensing their increasing ascendancy, the two boys had begun to ask for more and more money. Just a week ago, they had asked for an extortionate amount. When he had refused to pay so much, they had begun to make threats. Now, there was also the note from the girl and he was scared, really scared.

As it turned out, his fear was justified. The girl wasn't lying and, moreover, was determined to have the baby. When he refused to pay her blackmail demand, her father sued him for maintenance. He denied both paternity and ever having had sex with her, but was legally forced into a test of paternity. The publicity surrounding the case encouraged the two boys, now aged fourteen, to decide there was money to be made from selling their story. They accused the teacher and his lover of forcing them to take part in homosexual orgies against their will.

The teacher had judged his partner correctly. She stood by him during the early stages of the paternity case but left him when the homosexual scandal broke, taking their two daughters back to her own country. He never saw them again. Within a week of her departure, the paternity test confirmed him as the father of the schoolgirl's child. Following his arrest he was tried and found guilty of having sexual intercourse with minors, both male and female. He spent only a short time in jail, but it was long enough for him to contract HIV from homosexual activity with other inmates. Unemployed and penniless, he died of AIDS just before his thirty-seventh birthday.

~

Most readers of this book will be exclusively heterosexual. After a period of sexual exploration and mate selection during late adolescence, they will reproduce within the context of one or two successive long-term relationships. The men will have sex over their lifetime with about a dozen women – and the women with about eight men. On average, they will produce two children and eventually four grandchildren.

A minority, though, will pursue their reproductive success in a quite different way. A few will be bisexual, spending phases of their life directing most if not all of their sexual attention towards people of the same sex as themselves. Others will be very promiscuous, having hundreds if not thousands of sexual partners in their lifetime. Yet others, at the other end of the spectrum, will only ever have sex with one person. And there will be some men who will direct part of their sexual activity towards forcing women they have never met before to have sex with them. Some of these will band together in gangs before finding a woman to rape.

The conventional majority often find it difficult to understand the minority who show these alternative strategies. Their unusual behaviour is often interpreted as an aberration. The sometimes unpalatable truth of the matter is, however, that such minorities are pursuing reproductive success just as vigorously and strategically as the conventional majority. And we should not assume that just because these alternatives strategies are uncommon they are necessarily unsuccessful.

Each of the seven scenes in this chapter focuses on an area of human sexuality in which people pursue an uncommon but often successful reproductive strategy. In this first scene, we explore the way that male homosexual behaviour can be a successful reproductive alternative to exclusive heterosexuality.

Any discussion of homosexuality is bedevilled by the ambiguity of the words that are used. In the following pages, I shall use the following conventions. A *heterosexual* is a man who only ever has sex with women. An *exclusive homosexual* is a man who throughout his entire life only ever has sex with other men. A *bisexual* is a man who has sex with both men and women. *Homosexual behaviour,* on the other hand, is the behaviour shown towards other men – whether the man concerned is an exclusive homosexual *or* a bisexual.

At first sight, homosexual behaviour might seem a strange way to pursue reproductive success. It would seem particularly strange if, like most people, we made the mistake of assuming that just because a man is sexually attracted to other men, he is inevitably less likely to reproduce. The evidence, in fact, indicates the contrary. Far from being a pathway to a lower level of reproductive success, homosexual inclination is very much a successful reproductive alternative to heterosexuality.

Men who are attracted to other men do still reproduce – and on the whole they reproduce very successfully. On average, every person who reads this book will have had within the past five generations – in other words, since about 1875 – a male ancestor who practised homosexuality. This does not mean that we have all inherited a predisposition for homosexual behaviour. Some will have done, as we shall see, but only a minority. Nevertheless, it does mean that none of us would be the person we are today if one of our ancestors had not shown homosexual behaviour – and reproduced.

Before we start to discuss just how homosexual behaviour can aid a man in his pursuit of reproductive success, we must consider four basic facts about male homosexuality that are not generally known and that provide a most important perspective.

First, homosexual behaviour is not peculiar to humans. Adolescent birds and mammals often show such behaviour. Male monkeys show the same range of homosexual behaviour as men, from mutual caressing and masturbation to anal intercourse. There are reports, for example, of a male monkey masturbating to ejaculation while being penetrated anally by another male.

Secondly, as far as humans are concerned, homosexual behaviour is shown by only a minority of men – at least in the largest and most industrial societies. In Europe and the United States, for example, only about 6 per cent of men experience any homosexual contact during their lifetime, most often during adolescence. For two-thirds of those men, that contact is intimate and genital, often involving anal intercourse.

Thirdly, in all birds and mammals, including men, the vast majority of males who show homosexual behaviour are bisexual. For example, male monkeys who have anal intercourse with other males do not reduce their rate of intercourse with females. The same is generally true for men. The vast majority (80 per cent) of those who have sex with men also have sex with women. Many, like the man in Scene 30, may have phases that are exclusively or almost exclusively homosexual, but for fewer than 1 per cent of men does this 'phase' last an entire lifetime.

Finally, there is now convincing evidence that homosexual behaviour is inherited. Genetic inheritance is more often via the mother than the father: for example, men with homosexual inclinations are much more likely to have uncles and cousins with similar inclinations on their mother's side than on their father's. In the scene we have just witnessed, the man's uncle was more likely to have been his mother's brother than his father's; his cousin was more likely to have been the son of one his mother's brothers or sisters than his father's.

A genetic basis to homosexual behaviour does not mean that

the circumstances encountered by boys during their childhood do not also influence their behaviour. Boys genetically inclined towards homosexual behaviour may not show that inclination in some childhood situations but may do so in others. The man in the scene, who almost certainly carried the genes for homosexual behaviour, might never have developed his homosexual tendencies had it not been for his early relationship with his uncle. The converse could also be true, though it should be less common – boys without the genetic inclination may nevertheless be seduced or forced into homosexual behaviour during childhood. Modern evidence suggests that, more often than not, exclusive homosexuals and bisexuals are born and not made.

This discovery provides an important clue for biologists in their attempt to understand the evolution of homosexual behaviour. No gene can persist in a population at the 6 per cent level unless on average it imparts some reproductive advantage to the individual concerned. Of course, a lifetime of exclusive homosexuality can have no reproductive benefit – but bisexuality can. It seems most likely that exclusive homosexuality is a genetic by-product of the reproductively advantageous characteristic of bisexuality. If so, homosexual behaviour joins the ranks of a number of other human characteristics that are advantageous when a person has inherited a few of the relevant genes, but disadvantageous if they have inherited more.

The classic example of such a characteristic is sickle-cell anaemia. In the tropics, a single level of the sickle-cell gene is advantageous, endowing its possessor with increased resistance to malaria compared to people without the gene. A double level of the sickle-cell gene, however, condemns the possessor to an early death and/or a lifetime of pain and suffering.

Of course, this comparison between the genetics of homosexual behaviour and the genetics of sickle-cell anaemia should not be misinterpreted: there is no implication here that the former, too, is a disease. Rather, the anaemia is the best-studied example of a *genetic principle* that could also be applied to the inheritance of homosexual behaviour. We can think of bisexuals as having a small number of the genes for homosexual behaviour, and exclusive homosexuals as having a larger number – bisexuals have a reproductive advantage relative to heterosexuals; exclusive homosexuals never reproduce, and have a reproductive disadvantage compared with both heterosexuals and bisexuals.

So, how big is the advantage of bisexuality compared with a lifetime of only ever having sex with women?

As far as children within long-term partnerships are concerned, bisexual men have fewer children over their lifetime, but probably have them earlier in life. The man in Scene 30 had two children with his long-term partner, probably about average for the society in which he lived. But he had them before he was twenty-three, several years earlier than the average heterosexual. Such early reproduction may not seem very advantageous, but it can be. Biologists measure reproductive success not simply in terms of *number* of children or grandchildren, but in terms of *reproductive rate*. A person can have a higher reproductive rate than another either by producing more children over a lifetime or by producing the same number of children but earlier. Although throughout most of this book it has been adequate to discuss simply the pursuit of reproductive *success*, it is crucial here to remember that when we talk of reproductive success, we really mean reproductive *rate*.

It is difficult enough to compare the reproductive success of different categories of men, such as bisexuals and heterosexuals, even if we limit our comparison to long-term partnerships. The risk of including children raised within those partnerships but fathered by other men will always render such comparisons fragile. And our attempt becomes impossible when we try to compare their further success via short-term relationships with a number of women. Even the women who produce the children may not always know who the fathers are – so the men certainly won't. Yet it is precisely this avenue to reproductive success that seems to be the most important to bisexuals. The expectation is that such reproduction allows bisexuals to achieve a greater reproductive success than heterosexuals, but the evidence is impossible to obtain.

Multiple partners, both male and female, are a feature of male bisexuality. Nearly a quarter of men who show homosexual behaviour have more than ten male partners in a lifetime. For some the figure can be in the hundreds. More importantly, though, the more male partners a bisexual man has during his lifetime, the more female partners he is also likely to have. Since, on average, a bisexual man will inseminate more females over his lifetime than will a heterosexual man, a bisexual man is more likely to have children with different mothers.

The important question, of course, is whether the average bisexual's success at attracting and seducing many different women owes anything to his experiences with many different men. There are probably three main ways why it might.

The first is that early learning with other boys gives the bisexual a precocious sexual competence. Over 80 per cent of men who are ever going to show homosexual behaviour have done so by the time they are fifteen, and 98 per cent by the time they are twenty. Male homosexuality is an activity that occurs most often during adolescence or even childhood, whether with contemporaries or with older men. To appreciate the difference in competence between boys with homosexual experience and their heterosexual contemporaries, compare, for example, the man in Scene 30 with the man in Scene 27. The latter could scarcely manage intercourse, even at nineteen, let alone cope with the subtleties of the female orgasm. In contrast, even at thirteen the bisexual in Scene 30 could seduce girls into intercourse. By nineteen, women were queuing for his favours, and even in his mid-twenties his sexual aura made him attractive to girls from the age of fifteen upwards. The result was that, before his thirtieth birthday, he had fathered four children with three different women – more than most men in his society manage from a lifetime of heterosexuality.

The second way in which homosexual activity can aid heterosexual success is by allowing practice with different personalities. Experience with multiple male partners of different character types gives the bisexual an edge when interacting with multiple female partners of different character types (Scene 36). For example, the man in Scene 30 was aware of the similarity between his final male lover and one of the women with whom he had produced a child. Experience with one gave him experience at handling a relationship with the other. In this case, he experienced the woman before the man. When the reverse is true, experience gained with a man of a particular character type can help the bisexual to get the most out of a relationship with a woman of a similar character type. The experience may help at all stages and levels of that relationship – seduction, stimulation, social interaction and even deception.

The final way in which homosexual activity can aid heterosexual success is via infidelity from within a long-term heterosexual relationship. Practice at being unfaithful to his female partner with a man gives the bisexual experience at walking the tightrope of infidelity with a woman. Although bisexual men decrease their homosexual activity markedly as they leave adolescence and begin relationships with women, their homosexual inclination rarely disappears completely. A man with a long-term female partner is as secretive about his homosexual infidelity as he is about his heterosexual infidelity.

There is some advantage in practising infidelity with a man. The long-term female partner of a bisexual is less successful at detecting his homosexual infidelities than his heterosexual infidelities (often because she does not know he is bisexual). A woman who does not know her partner's true sexuality is likely to assume that he is heterosexual – because the majority of men are. Consequently, she will usually feel less threatened by his relationships with other men than by his relationships with women – the *average* man's relationships with other men will be less likely to be sexual than his relationships with women. Even on those occasions when her partner's relationships with men are sexual, a woman has less to lose, at least initially, than if he is unfaithful with a woman. Although some of the costs of infidelity (Scenes 9 and 11), such as risk of infection, still apply, most do not. For example, he will never need to reduce his support for *her* in order to help maintain his lover's child. He is probably also less likely to desert her to live with his lover if the latter is a man.

So homosexual behaviour during adolescence and beyond can give a man considerable reproductive advantages over his heterosexual contemporaries. In which case, why is bisexuality not more common?

The answer is relatively straightforward. There are costs to bisexuality which can negate the benefits. The most important cost of homosexual behaviour is a greater risk of disease. Even before the advent of AIDS, homosexual behaviour brought with it an increased risk of early death from sexually transmitted diseases, such as syphilis. In effect, bisexuals are programmed by their genes to pursue a lifestyle that trades the benefit of an earlier and perhaps greater production of children (with more women) against the risk of an early death.

Another cost is genetic – and here there is another parallel with sickle-cell anaemia. Although people with a small number of the genes may gain an immediate apparent advantage, as we have seen, that advantage may not be as great as it seems. This is because, compared with people without the genes at all, those with a small number will produce descendants with a higher proportion of individuals with a larger number. In other words, although bisexuals produce more children and grandchildren at a faster rate than heterosexuals, among those descendants there will be a few exclusive homosexuals who in turn fail to reproduce at all.

Yet another cost arises because a proportion of the heterosexual majority display homophobia – a prejudice against people who show homosexual behaviour. Such prejudice is occasionally so extreme and violent that any man suspected of homosexual behaviour faces an increased risk of injury or even death. We encountered a similar if less extreme prejudice in Scenes 12 and 13 in relation to masturbation. With masturbation, of course, such prejudice is all bluff and hypocrisy – the intimidators being as likely to masturbate as the intimidated. Inevitably, some homophobes are also hypocrites, displaying public homophobia while secretly behaving bisexually. On the whole, though, most homophobes are part of the heterosexual majority.

Whenever we encounter such common prejudice, it is usually because the target is in some way a threat to the people showing the prejudice. It is quite possible that homophobes, like bisexuals, are themselves born not made – an inevitable evolutionary result of the very success of bisexuality that we have just discussed. The reproductive advantages enjoyed by bisexuals mean that they *should* – for that reason alone – be regarded as a threat by the surrounding heterosexuals. Unfortunately, the bisexual's role in the spread of disease adds to this threat. So, just as we discussed for masturbation (Scene 13), one defence for the surrounding individuals is to try to reduce the bisexuals' reproductive advantages through threat and intimidation.

The final picture, therefore, is that compared with their heterosexual contemporaries, bisexuals experience both advantages and disadvantages in their pursuit of reproductive success. In which case, the important question becomes whether the total costs are greater than the total benefits, or vice versa. Are bisexuals more successful reproductively than heterosexuals – or are they less successful? The answer is that it depends on how common bisexuals are in the population. When they are rare, they are more successful than heterosexuals. When they are common, they are less successful. The reasons are as follows.

The advantage of bisexuality lies in its exponents potentially having a higher *rate* of reproduction than the average for the society in which they live. As we have noted, earlier and better learning of sexual technique gives bisexuals a competitive advantage over other men in gaining sexual access to women. The more bisexuals there are in the population, however, the greater the chance that their competitors are also bisexual – and the more common bisexuals become, the less the advantage enjoyed by any one individual as a result of his bisexuality.

As the proportion of bisexuals in the population increases, not only does the advantage of bisexuality decrease, but the costs increase. Of the three costs discussed above, two – genetics and disease – clearly increase as bisexuality becomes more common.

As far as the genetic risk is concerned, the more common the genes for homosexual behaviour the greater the chance that any two people will carry those genes, and thus the greater the chance that a man and his partner will produce sons and grandsons who are exclusively homosexual and will therefore fail to reproduce at all. When it comes to the risk of disease, the greater the level of homosexual activity in the population, the faster the spread of disease. More people, both heterosexuals and bisexuals, will become infected. But because bisexuals are always at greater risk of contracting disease, they will suffer most. Thus, the chances that any given bisexual will have an early death are greater.

We have seen that, in a population in which bisexuals are rare, those who are bisexual enjoy considerable advantages over heterosexuals. The result is that the genes for bisexuality increase in proportion in the population. Conversely, as bisexual behaviour becomes more common, the advantage enjoyed by bisexual individuals decreases and the costs increase. If the behaviour becomes *too* common, the reproductive rate of bisexuals falls below that of heterosexuals, and the proportion of bisexuals in the population begins to decrease once more.

The inevitable result of this interplay of costs and benefits as the proportion of bisexuals rises and falls is that eventually the proportion of bisexuals stabilises. Moreover, it stabilises at precisely the level at which the average success of bisexual behaviour in each generation is just the same as the average success of heterosexual behaviour. Therefore, the answer to our question – who does better reproductively, bisexuals or heterosexuals? – is 'neither'. The only difference between the two is that the reproductive success of bisexuals is more precarious – they have a much greater chance of not reproducing at all. However, if they successfully avoid being killed by homophobes and contracting HIV, they also have the *potential* to be very successful. On average, the greater risks and the greater potential just balance each other.

The conclusion, therefore, is that in the larger, more industrial societies the genes for bisexuality have stabilised at around 6 per cent of the population because this is the level at which bisexual and heterosexual men, on average, do equally well.

The situation would be very different, of course, if the costs of bisexuality were never as great as the benefits, no matter how common the bisexual genes became in the population. Suppose, for example, there were some societies in which there was very little risk of sexually transmitted disease. In such societies, the benefits of bisexuality could always outweigh the costs, no matter how many people used the strategy. We should expect the genes for bisexuality to sweep through the whole population. Is it conceivable that there have ever been societies in which there were sufficiently few costs to bisexuality? The answer is yes.

So far in this discussion, we have concentrated on large industrialised societies – which are uniquely unfavourable for the wholesale spread and persistence of bisexuality. In particular, they harbour and facilitate the spread of those sexually transmitted diseases which generate the main costs of the strategy. The recent emergence and spread of HIV and AIDS are merely the most recent examples of a sequence of events that will have occurred many times throughout human history. For example, from the Middle Ages until the twentieth century syphilis was the major sexually transmitted killer in the larger societies.

Historically, small, more isolated communities have harboured relatively few diseases. Since the members of such communities were the offspring of the survivors of past epidemics, they inherited their ancestors' natural, perhaps genetic, immunity. New diseases rarely appeared because the people had little contact with the outside world. When they did have such contact, few escaped exposure to disease no matter what their behaviour. The survivors were again those with some form of immunity – which they passed on to their offspring. Consequently, before they encountered the outside world and were exposed to measles, smallpox, syphilis and now AIDS, such small isolated societies enjoyed long periods with little danger from disease. In such circumstances, bisexuality would have carried nowhere near the risks that it now does in larger societies, with the result that the genes for bisexuality should have spread unchecked by disease. No matter how common bisexuality became, its exponents would still have reproduced faster than heterosexuals. It should be no surprise, therefore, to find that many such communities when first encountered and studied contained a far higher proportion of bisexuals than the industrialised world. And it should be no surprise, either, to find that, when the majority of people are bisexual, homophobia either becomes much less common or disappears completely.

As far as the level and tolerance of bisexuality are concerned, it is the major industrialised societies that are the exception, not the rule. Anthropologically, in 60 per cent of human societies bisexuality is both common and socially accepted. Some societies, such as certain small island communities in Melanesia, accept as normal that *all* adolescent males will at some point engage in homosexual anal intercourse. The women also accept that their long-term partner will from time to time have sex with other men, and they tolerate their partner's homosexual infidelity more than his heterosexual infidelity. The usual attitude is that their partners can continue with their homosexual activity as long as it does not interfere with their heterosexual relationship. But even in these societies, where all male adolescents have phases of homosexuality, sometimes within short-term 'monogamous' relationships, exclusive homosexuality over an entire lifetime is very rare. Homosexual behaviour is very clearly part of a bisexual reproductive strategy. Moreover, it is so successful a strategy that it has completely displaced the heterosexuality that is the norm in larger, more disease-prone societies.

SCENE 31

The Coming of Women

The girl rubbed the steam off the bathroom mirror so that she could examine her face yet again. Reassuring herself for the fourth time that evening that the spot on her chin had almost gone, she carried on towelling herself dry. Twenty years old and she was still getting spots – surely she must grow out of them soon? As she hung the towel over the radiator, she noticed the nearly empty packet of tampons on the shelf and smiled. Which one of them would make the supreme effort and put it back in the cabinet this month? It was a week since their periods had finished – strange how they tended to coincide – and neither of them had yet managed to put the packet away. Powdering herself, she felt her throat tightening. Even after a year together she still got excited at the prospect of sex with her partner. She had been looking forward to this all day and there was a tingle between her legs already.

As she walked naked from the steamy bathroom to the dry heat of the bedroom, her partner was already lying on the bed. Ten years older than herself and the mother of a young child, she still had a wonderful body. How much better it was to lie down next to a soft, smooth, yielding and compliant woman than a muscular, urgent and selfish man.

Immediately she was on the bed, they embraced and kissed. As they did so, their hands slid expertly over each other's bodies, lightly stroking here, lightly caressing there. From time to time one would massage the other's breasts, play with a nipple, or gently stroke pubic hair. Taking it in turns, they moved down each other's bodies, kissing and licking as they went, each doing to the other what she most wanted to be done to herself. After a while, the girl sat astride her partner, facing away from her, then bent forward to lick her thighs and genitals. As she did so, she raised her bottom in the air, presenting herself to her partner's tongue. This was the bit she liked best. Whenever she felt like an orgasm, this never failed to take her to the brink. As the warm, wet tongue licked and probed her vaginal lips, then so gently massaged her clitoris, she felt the sensations she had been anticipating all day. They kept each other on a high for minutes as they neared the edge but never crossed. Exquisite though they both found oral sex, they rarely climaxed from that alone.

Finally, unable to wait for release any longer, she lay back alongside her partner. Her momentary loss of focus soon passed as they kissed deeply, each briefly tasting their own juices on the other's tongue. As they kissed, their fingers stroked and probed each other's wet vaginal lips, then massaged each other's clitoris. Each knew exactly what to do. As the girl felt the flush spread over her chest, throat and face she saw the same happening to her lover. Their breathing became a gasping, their pulses raced, and the sounds in their throats became louder and more urgent. For one brief, wonderful moment they hovered on the brink, then climaxed more or less simultaneously. It had easily lived up to the day's anticipation. They settled into their usual post-orgasmic embrace, hands now idly stroking each other's body. In the seconds before drifting off to sleep, she thought how much more expert they had become at giving each other orgasms since they first started living together.

Normally, they slept for only fifteen minutes or so after sex, but when she woke nearly an hour had passed. Her partner was still asleep and it seemed cruel to wake her. Nevertheless, she was beginning to fret. She had only an hour to get dressed and cross town. Eventually, she could delay no longer and made to move off the bed. Immediately, her lover woke and sleepily urged her to stay a little longer, suggesting it wouldn't matter if she was a bit late for her meeting. Sliding off the bed and going in search of clothes, the girl said she wished she could stay but really did not want to be late for the third week in a row.

As she went into the bathroom, her partner shouted after her, complaining that she always seemed to be out in the evenings these days. She reminded her that in the beginning they were never out of each other's company. That had been a year ago, the girl shouted in reply. Coming back into the room, she reassured her companion that once her exams were over, they could return to how they had been. In the meantime, she couldn't afford to miss anything.

Before she left, her lover made her promise she wouldn't be late. Saying that she would do her best, but warning that she would probably be dragged off for a drink once the meeting was over, the girl left.

Once alone, the woman got up, still naked, and switched on some music. Idly, she walked round the room. A moment of fear struck her when she saw the letter on the shelf, reminding her to have a repeat cervical smear tomorrow – abnormal cells, was that serious? In an attempt at self-distraction, she picked up the framed photograph of herself and the girl on holiday the previous summer, both tanned, drunk and happy. Not far away on the shelf was a photograph of herself with her young son. He was now ten years old, but it was several months since she had seen him. Once she had begun living openly with her current lover, his father had refused her access, insisting on looking after the boy himself with his new partner. If only he knew, she thought, that the boy almost certainly wasn't his but, most likely, the product of a wild evening she had spent in a hotel room with two men and a girl while he had been away. Memories of that evening triggered the familiar tingle between her legs. It was only an hour since her last climax, but the minute she felt that sensation she knew she would masturbate before the evening was over.

The girl got off the bus and walked into the restaurant. She was late. She had known it would be a mistake to try to fit in sex before leaving but had been desperate for an orgasm all day. Looking around the tables, she wondered if he would still be there. This was their first real date, though he had been pushing her for some time. She didn't really want to deceive her partner, especially now. The latter's problems over access to her son and the worry of her abnormal cervical smear results were putting them both under considerable strain. Somehow, though, she didn't feel quite as guilty about a clandestine meeting with a man as she would with another woman.

It was over a year since she had had sex with a man. In fact, her current lesbian relationship had been so rewarding that she had begun to think she would never feel like sex with a man again. She had always known she was bisexual, ever since she was a child. Throughout childhood and adolescence she had persuaded girlfriend after girlfriend to take off their clothes and get into bed with her. Once in her mid-teens, however, she had also given in to the occasional male.

At first she had just been curious about heterosexual sex, but after a few experiences had actually begun to enjoy it, in its own way. More than anything, though, she had enjoyed the feeling of power she could have over boys. Compared with her girlfriends, first the boys and later the men had seemed so gullible, so easy to manipulate and deceive, that she had found it difficult to respect them. Also, men were so sexually inept; so selfish. As long as she was in the mood, she could virtually guarantee that another woman could give her an orgasm. With a man, she nearly always had to apply the final touches herself, if she felt like it. So what was she doing in this restaurant, clandestinely meeting a man? She really didn't know, except that perhaps she was ready for another experience – and this man really was rather attractive. A hand waved from the corner table. She waved back and went to join the hand's owner.

On the other side of town, while the pair in the restaurant ate, the girl's lover was lying naked on the bed, bringing herself to a climax. Afterwards, she went into the sitting-room to relax with a bottle of wine, music and a book. But she couldn't settle. There had been something strange about her lover's manner for the past day or so and it was bothering her. Suddenly, she began to look up the telephone number of the bar where the girl and her friends usually went after their lectures and meetings. She phoned, asking to speak to her, intending to suggest that she came and joined them. It wouldn't be the first time. She was told, however, that neither the girl nor her friends were there. Deciding they must have gone elsewhere, she tried to settle down to her book once more.

There was an inquisition when the girl eventually returned – late, but not unduly so. The story she gave was that after their meeting her friends had decided they should all try somewhere new. Eventually, she convinced her partner she was telling the truth, but there was still an atmosphere between them when they went to bed.

In fact, the girl and her male escort had not had sex that

evening, but it was only a week before they did. Afterwards, as the weeks passed, their relationship developed in a way that she had only thought possible with a woman. Her infidelity to her lesbian partner became increasingly difficult to hide as she found herself wanting to spend more and more time with her male lover. Her ingenuity was stretched to the limit. She also began to feel guilty at what more in the way of trouble she might be giving to a woman whose life was becoming increasingly fraught in every way. Her repeat smear again indicated abnormal cells – and she began the stressful process of having further tests and waiting for results. As if that wasn't bad enough, she continued to be denied access to her son and was passed over for promotion at work when the rumour that she was a lesbian, which had been in the air for a while, began to spread.

The woman had begun to suspect that her young lover was being unfaithful to her, and increasingly they began to argue. Their worst fight was sparked off when they accused each other of faking orgasms. But she still couldn't be certain about her partner's infidelity – right up to the moment, that is, that the latter broke the news of her pregnancy. At first, she tried to convince her that they could look after the baby together, but the girl was adamant about moving out and living with the baby's father.

For a time, the woman was ill with stress and loneliness. After losing her job, she began to feel suicidal. But then, in the depths of depression, she met a man who had just been deserted by his partner and family. They counselled each other out of their joint despair and within weeks were living together. A year later, she had his child – a daughter. Not long afterwards she was operated on successfully for cervical cancer, and from then on her life improved.

Once her son became independent of his father, he came to see her. Untainted by his father's homophobia, he quickly grew close to his mother and his younger half-sister and they became very much a family. Although the woman died relatively young, in her sixties, it was not before she had enjoyed fifteen contented years as a doting grandmother.

~

Having discussed the way that homosexual behaviour can aid men in their pursuit of reproductive success (Scene 30), it is relatively straightforward to do the same here for women. This is because there are many similarities between male and female bisexuality and relatively few differences.

Even those differences that *do* exist are largely just a matter of degree. For example, on average across societies there are fewer bisexual females than males – a trend that is also shown in most animals. In any given human society, there tend to be about a third to half as many women bisexuals as men. In large industrial societies, in which only about 6 per cent of men show homosexual behaviour, about 2–3 per cent of women do so. In societies in which all men show homosexual behaviour, about 30–50 per cent of women do so. This difference means that on average we should each need to go slightly further back in our family tree to find a female bisexual than to find a male. But the difference is only about one generation – back to about 1850 instead of 1875.

Not only are there fewer female bisexuals than male, but on average they begin their homosexual (lesbian) activities a little later in life. Only 50 per cent of bisexual women have had their first homosexual experience by the age of twenty-five, and only 77 per cent by the age of thirty. Some do not have their first lesbian experience until they are in their forties.

Another difference between male and female bisexuals is that bisexual women do not have as many homosexual partners as bisexual men. Only 4 per cent of bisexual women have more than ten homosexual partners in their lifetime compared with 22 per cent of bisexual men. Similarly, women are more likely to have longer-lasting 'monogamous' relationships with each other than are men. A common pattern, shown by the girl in the scene, is for a woman to stay in a homosexual relationship for one to three years before moving on to a heterosexual relationship. Also as in the scene, older women often 'fit in' a stable homosexual relationship between successive heterosexual ones.

Compared with these minor differences between male and female homosexual behaviour, there are many important similarities. For example, most women who show homosexual behaviour are bisexual. Fewer than 1 per cent of women in any society are *exclusively* homosexual throughout their lifetime. Over 80 per cent who show lesbian behaviour also show heterosexual behaviour, as did both women in Scene 31. All human societies contain bisexual women; moreover, female bisexuality is genetic and inheritable. And, like all of the characteristics we have discussed, female bisexuality is widespread among different species of mammals, birds and reptiles. In fact, there is one species of lizard which consists only of females. These will not develop eggs unless they are first mounted in a pseudo-copulatory act by another female. They take it in turns to mount each other and to stimulate each other's egg production.

The many similarities and few differences between male and female bisexuality are significant enough to justify a similar interpretation of the two types of behaviour. When we compare the reproductive success of female bisexuals and heterosexuals, we reach precisely the same conclusion as we did for men (Scene 30) – as a strategy for the pursuit of reproductive success, female bisexuality is a real and successful alternative to heterosexuality.

Briefly, bisexual women reproduce earlier than heterosexuals but are also more at risk to disease and an early curtailment of their reproductive life. Just as for men, the net advantage of bisexuality tends to be greatest when the behaviour is rare. The proportion of women in any given population who are bisexual, therefore, should reflect the level at which the costs and benefits just balance each other.

In many ways, we know more about the costs and benefits of female bisexuality. This is because it is much easier to know how many children a woman has had at different stages in her life than it is a man – especially a man who has had many different partners. For example, we know that by the age of twenty a bisexual woman is four times more likely than a heterosexual to have a child, and that even by the age of twenty-five she is still twice as likely. The woman in Scene 31 had had one child by the time she was twenty and two by the age of thirty-one. By the end of their reproductive lives, however, bisexual women tend to have fewer children than heterosexual. In Britain in the 1980s, for example, one survey gave the average figures to be 1.6 children for bisexuals and 2.2 for heterosexuals. The trends for earlier but fewer children for bisexuals just cancel out, giving them the same overall *rate* of reproduction as heterosexuals.

As briefly mentioned above, part of the reason that bisexuals often end up with fewer children despite an earlier start than heterosexuals is that, like the woman in the scene, their reproductive life can be curtailed through disease. Just like male bisexuals, females are at greater risk of contracting sexually transmitted diseases. By the age of twenty, they are more likely to have experienced genital infections. By the age of twenty-five they are more likely to exhibit abnormal cells in cervical smear tests, and by thirty they are more likely to have cervical cancer.

To what extent this increased risk of disease is a direct result of bisexual activity is not known. Sexually transmitted diseases, such as herpes and genital warts, can be passed directly from woman to woman during love-making. But there are other aspects of female bisexuality which could also increase the risk of infection, as we shall soon see.

The fact that female bisexuality is less common than male suggests to an evolutionary biologist either that women have less to gain from bisexuality than men or that they have more to lose – or both. However, it is unlikely that bisexual women have more to lose – quite the opposite, in fact. For example, not only do male bisexuals seem much more likely to be at risk to disease, they are much more at risk to violence from homophobes than are female bisexuals. This is probably because lesbians are much less of a reproductive threat to heterosexual women than are gays to heterosexual men.

The conclusion, therefore, is that if women do not have more to lose from bisexuality than men, they must have less to gain – and this should not be so surprising. We have already seen that males have a lot more to learn about sexual technique during early adolescence than females (Scenes 27 to 30). Homosexual experience at the appropriate time can confer precocious sexual competence on both sexes, but does so much more clearly on males. In fact, given that females have fewer basic sexual techniques to learn than males during adolescence (Scene 27), we have to ask what it is about sexual experience with other women that gives bisexuals such an early advantage over heterosexuals.

The main things that a woman has to learn about sex are those techniques that allow her to get the most out of long-term relationships (Scene 18). In particular she needs to learn the techniques of infidelity and deception. She also needs to learn how to make the most of her orgasms in a way that will give her the greatest control over sperm retention and warfare (Scene 26).

The elements of deception needed to exploit infidelity are probably *better* practised with another woman than with a man. Moreover, they are best practised within the context of a relatively long-term monogamous relationship with a woman – which is one of the key features of female homosexual behaviour. If a woman can fool a long-term female partner over infidelity, faked orgasms and the like, then she should find it relatively easy to fool a similar male partner. This allows her to exploit the advantages of infidelity and sperm warfare (Scenes 21 and 26) while reducing the risk of having to pay the costs (Scenes 9 and 11).

If bisexual women learn earlier and better how to exploit relationships and how to promote and influence sperm warfare, it should not be surprising to find that they make earlier and greater use of these abilities. Bisexual women are inseminated just as many times over their lifetime by a man as are heterosexual women. Moreover, bisexual women are more likely to have more than one concurrent male partner, and to have sex with two different men in a short enough period of time to promote sperm warfare.

Not only do bisexual women promote sperm warfare more often, they seem better able (though the finer details have not been measured) to control sperm retention and thus the outcome of that warfare. They are more likely to masturbate, and to masturbate more often, than heterosexual women. They thus create stronger cervical filters (Scene 22). When they do have sex with a man, bisexuals are no more nor less likely to have an orgasm. But they are more likely to have a bypass orgasm (Scene 25), negating the effect of their cervical filter.

Because many of the orgasm responses that are a woman's greatest weapon in sperm warfare actually occur in the absence of men anyway (masturbatory; nocturnal – Scenes 22 to 26), they can easily be practised in the context of a lesbian relationship – including even those orgasms that in a heterosexual relationship normally occur before, during or after penetration. This is because the stimuli women give each other during lesbian sex are very similar to those given by a man during foreplay. The most common technique used by lesbians is the stroking and massaging of the genitals, particularly the clitoris. After that, in decreasing order of frequency, come the stroking and massaging of the breasts, the licking and sucking of nipples, oral sex, and the pressing and rubbing together of the genitals. In societies such as some in Melanesia, lesbians only ever use their mouths and hands during sex. Elsewhere, objects are sometimes used, either to stimulate the clitoris or to insert in the vagina. These range from reindeer calf muscle in parts of Siberia to bananas, sweet potatoes, and elegant carved wooden phalluses in other parts of the world. In the more industrialised societies, of course, commercially available vibrators and dildoes are sometimes used. However, penetration either by fingers or objects is still a relatively infrequent part of lesbian stimulation. In one American survey, only 3 per cent of lesbians said that they regularly used penetration to stimulate their partner during sex.

The orgasm 'success rate' is about twice as great if a woman is stimulated by another woman rather than by a man. Moreover, the orgasms women give each other are much more likely to occur during the fertile phase of their cycle. The peak of sexual activity that we witnessed right at the beginning of Scene 31 took place a week after the women had finished their periods. If they were going to ovulate that month (and they might not – Scene 15) it would be within the next few days. They were both in their fertile phase.

Interestingly, this peak of orgasm frequency during the fertile phase of a lesbian's menstrual cycle is mirrored by peaks of homosexual activity by other female animals at a similar phase of their cycle. Whether we study cows, rats or guinea-pigs, we find that females are much more likely to mount each other during their fertile phase. It has been shown experimentally that such mounting behaviour is hormone-linked, so that changing the hormone balance during the menstrual cycle alters the timing of mounting, and can even stop it altogether. Lesbians who take oral contraceptives lose their peak of mid-cycle orgasm activity. This suggests that, like the other animals described above, a lesbian's motivation to have orgasms with her partner is also under hormonal rather than cerebral control.

Our final picture of the homosexual behaviour of women, then, shows a process of practising with women for success at getting the most out of longer-term relationships with individual men. This is not really any different from our conclusion for the advantage of bisexuality to men. Bisexual men, however, are practising primarily for success with many women. Unlike a man, a woman cannot increase her reproductive success simply by having more sex with more men. She does so, instead, by being much more selective and strategic and by thus making the most of exploiting the men around her. In this, a bisexual woman is aided by a precocious ability to promote and manipulate sperm warfare and by a longer-term ability to make better use of infidelity. Just as for male bisexuals, however, the reproductive strategy of female bisexuals brings with it increased risks that can result in an early curtailment of their reproductive activity through disease or early death. The net result, therefore, is that female bisexuals and heterosexuals have similar reproductive success – but achieve it by different means.

There is one last feature of interest in Scene 31 that we have not yet discussed. When women live together, they often synchronise their periods. Not only lesbians, but mothers and daughters, nuns, prisoners, nurses and students also often synchronise when they live together. In a fascinating series of experiments carried out in the USA in the early 1980s, each woman in a group of volunteers agreed to have the underarm secretions from another woman rubbed under her nose every other day for a few months. Each woman who received the secretion adjusted her menstrual cycle to synchronise with that of the woman who had provided it – which implies that chemicals are present in a woman's underarm secretions which enable groups who spend a lot of time together to synchronise their menstrual cycles.

In recent years, there has been a lot of discussion over whether menstrual synchrony is a real phenomenon. The most recent research suggests that it is – but that so too is *desynchrony*. In some groups of women, the menstrual cycles become more similar; in others, more different. Which direction is taken by any particular group is not due to chance, but depends more than anything on how many of the women in the group are ovulating regularly (Scene 15). Groups in which few are ovulating, particularly if some are on the pill, tend to synchronise. Groups in which most are ovulating tend to desynchronise. It is as if the women's bodies are trying to ovulate as many days apart from each other as possible.

Just why these responses should exist is still a mystery, but it probably has something to do with hiding the fertile phase from men (Scene 2). If a group of women all ovulated at the same time, even the most unaware of males might notice the behavioural changes associated with the fertile phase (Scenes 3, 6, 10 and 22). In isolation, an individual can more easily hide these changes from her partner amidst random swings in mood and behaviour.

The fact that the two women in Scene 31 were synchronised suggests that they had not ovulated in recent months. This is a common response to the absence of men (Scene 15). With the beginning of the younger girl's heterosexual activity, we should expect her first to ovulate, then to desynchronise her cycle from her female lover.

SCENE 32

The Tenth Tonight

The door-bell rang. It was uncanny the way she could tell something about the person at the door just from the way he rang the bell. The previous ring had been aggressive and persistent, but this one had a tentativeness about it. She got up off the bed and pulled on her bath robe over her underwear. Young, she thought to herself – or a vicar.

On her way down the stairs, she did some quick mental arithmetic. It had been a busy night. This would be the tenth – maybe she should also make it the last.

When she saw him standing there, she smiled to herself – he was young, not a vicar. And he looked distinctly uncomfortable. Diffidently, he asked her how much. She told him, then pointed out that it was more if he didn't want to use a condom. Taken aback partly by the price and partly by her appearance, he hesitated, then agreed. She stood back to let him in, told him to go on up, and closed the door. Before following him, she took her 'Model – first floor' notice out of the window.

Once in the room, she asked him which it was to be, condom or no condom. He said it would have to be with a condom because he didn't have enough money to pay for it without. 'Money first,' she told him.

The youth looked very nervous. To settle him she told him her name, or rather a name, and asked for his. He told her his family name. Despite herself, she laughed. She asked him if he was in the army. Embarrassed, he explained he was a student, that he had only just left school and couldn't get out of the habit of giving his family name when asked. She told him to relax, then asked if it was his first time. When he said it was, she asked if it was his first time ever or just the first time he'd paid for it. It was the first time he had paid and might be the first time ever, he said. He wasn't sure. He had got close a couple of times but didn't know if he really had or not.

She didn't ask him to elaborate – just told him to relax and leave it to her. Then she took off her bath robe and knickers, lay down on the bed, and told him to take off his trousers and pants and join her. She didn't react at all when she saw that his penis was limp and small, as if it had been hiding under his clothes. With difficulty, she put a condom on him, then did her best to encourage an erection. It was many minutes before she felt any reaction on his part, despite her soothing talk and practised hand. Then just as he began to stiffen and she thought she might be able to help him after all, he ejaculated, long before she could try to put him inside her. His look of dejection and embarrassment touched her. She told him not to worry and, as he hurriedly dressed to leave the scene of such humiliation, tried to reassure him that he would be fine next time. It was just nerves. She saw it happen all the time, she said.

After he had gone, she dressed. As she reckoned up her night's earnings, five with a condom and five without, the thought crossed her mind that maybe she should become a sex therapist when she retired. A few minutes later she was walking out of her front door and on to the street. Hailing a taxi, she set off for home.

As she watched the familiar route unfurl, her thoughts drifted back to the student and she marvelled at the way education seemed to slow down a person's sexual development. Not that being a student had hindered her. In fact, it was when she was at college, about fifteen years ago, that she had started on her career. Nobody could deny it had been profitable, even if it hadn't been the career she had envisaged when she left home.

Short of money, as were many of her fellow students, she had exploited her looks and joined an escort agency. Tempted by the financial offers she received, she had begun to spend the occasional night with those clients she found most attractive. By the time she graduated, she was earning so much money from sex, had accumulated such a large clientele of her own and, frankly, was enjoying herself so much that she saw no point in embarking on any other career.

At one time or another during her fifteen years she had exploited just about every avenue by which an attractive girl could make money from sex. A chance encounter gave her a brief spell making hard-core pornographic films. In the main, it had been fun. She laughed to herself as she remembered the contortions necessary, amidst a dozen naked men, to get a penis in both hands and all three orifices and still leave room for a camera. Even so, when a rich politician from among her clients decided he wanted exclusive access and set her up in an expensive flat in the heart of the city, she had retired from her film career to become a professional mistress. She didn't entirely give up having sex with other men, but pretended she had – and got away with it. Eventually, the politician's sex life had been exposed by a tabloid newspaper and she moved on – this time to a judge.

After the judge died, conversations with some of her fellow professionals led her into trying the streets for a while. In many ways, she had enjoyed her time on the streets more than any other phase of her career. It had certainly paid well. Eventually, however, she gave it up. After a few close shaves, a bad experience with a particularly violent customer had put her in hospital for a few days. After that, she had moved into the relative security of a bordello, which she also enjoyed. She only left when her looks began to fade and she could no longer compete in open parade with the younger girls. At least with her 'Model – first floor' method of recruitment, men had already half committed themselves by the time she opened the door. Anyway, she was still attractive enough for most to accept her price rather than walk away when they saw her.

She was going to retire soon. She knew she had been lucky and that she should quit while she was winning. Despite several thousand sexual partners in her lifetime, she had only had three bouts of disease and all had cleared up quickly with the help of antibiotics. She had been beaten up a few times, but no serious damage had been done. Most importantly, though, she had managed to avoid the drugs that had been, and still were, the ruination of so many of her fellow prostitutes. For all these years she had been making as much money each night as most people make in a week, or even a month. There had been overheads, of course – protectors to pay and places to rent. But even so, she could have retired years ago and still have lived comfortably off the interest from her savings. And the reason she hadn't retired was simply that she enjoyed her job too much. She would be loath to leave it all behind. In fact, she was seriously considering starting her own escort agency or bordello when she did eventually decide it was time to stop.

The taxi pulled up outside her home – a large detached house on the outskirts of town. Her partner shouted from the kitchen that it was good timing – the dinner was nearly ready and the children were already in bed. She went upstairs, douched herself, then had a bath. While she lay back in the warm water, her partner brought her a glass of wine, then went back downstairs to finish the cooking. Before going down to join him, she looked in on her four children, each in their own bedroom. It would be interesting to know who their fathers were. She liked to think that the two daughters, now ten and eight, were fathered by the politician and the judge, but she couldn't be certain. Her elder son was conceived while she was in the bordello, and could be anybody's. Many of the clients were men of wealth and status and there were several she wouldn't have minded being his father. Her younger son should be her partner's. She gave up working for a while specifically to conceive his child, but she became pregnant so quickly they still couldn't be entirely certain he was the father.

Her partner was five years younger than her. An ex-student and an ex-client, he had volunteered to look after the children in exchange for his keep and sexual access. They had been together now for five years.

After the meal the couple drank, and talked. For the first time that week, they had sex before going to bed.

～

Biologically, a prostitute is an individual who offers other individuals sexual access in exchange for one or more resources. In humans, the resource sought and offered is usually money, but it might just as well be food, shelter or protection. Nowhere is sperm warfare more rife than in the reproductive tract of active female prostitutes. By the end of a night's work, the number of sperm armies waging war inside a prostitute's body often runs into double figures. And, from time to time, the winner of that war claims the prize of fertilisation.

Prostitutes reproduce. Often, like the woman in the scene, they reproduce very successfully. Why do some women abandon fidelity, or secret infidelity, and openly pursue their reproductive success via prostitution? When they do, how does their reproductive success compare with that of women who pursue more conventional strategies?

Female prostitutes have been, and are, an almost universal feature of human societies. Anthropologically, only 4 per cent of societies claim not to contain any. The remainder acknowledge their presence. It is difficult, though, even in these societies, to estimate what proportion of women engage in prostitution at some time during their lives. Estimates of the number of overt prostitutes active at any one time range from less than 1 per cent of women in Britain in the late 1980s to about 25 per cent of women in Addis Ababa, Ethiopia, in 1974. Such estimates, however, are unreliable and are probably underestimates. More women than this will *sometimes* engage in prostitution.

Part of the problem lies in the lack of a precise definition. Overt, promiscuous sex in exchange for money is, in many ways, simply the least ambiguous exchange of sexual access for resources or 'gifts'. Throughout human history and culture, there are examples of men giving the woman (or her family) a 'gift' around the time of their first intercourse, without the woman being categorised as a prostitute. Often the exchange is ritualised, as during marriage ceremonies. Even on the wedding night, money may be demanded by the woman or her family before intercourse is allowed to take place.

Clearly, there are degrees of prostitution. In principle it is difficult to know where to draw the line between a traditional prostitute exchanging insemination for money and a woman in a long-term partnership exchanging insemination for support, protection and 'gifts'. The woman in this scene is clearly a prostitute, but what about the woman in Scene 18?

It is similarly difficult to decide for other animals just what is prostitution and what is not. At one extreme, there are clear acts of prostitution – like the female empid fly who trades sex for food. For a male empid to be given the opportunity of mating, he first has to find a swarm of gnats, catch one, wrap it in silk from his salivary glands, then find a waiting female and offer it to her. While she unwraps and then eats her meal, he is allowed to mate. The larger the gift, the longer it takes her to feed, the longer the male is allowed to mate, the more sperm he transfers, and the more eggs he fertilises. Once he has gone, the female waits for the next male to bring her food and to mate. In some species, females are so successful as prostitutes that they never need to find food themselves.

Migratory birds occupy a position at the other extreme of the prostitution scale. Males return to the breeding areas first and compete for the best territories – those in which they and a female can most successfully raise their young. Females arrive later and inspect the different territories and the males who defend them. Each female then makes her choice of the best compromise (Scene 18) between territory quality, male quality and availability (because the best territories and males are quickly snapped up by the earliest females). Eventually, a female will allow a particular male to have sex with her. In exchange she is allowed to share his territory. If the male is ousted by another, the female does not leave with her former partner, but allows the new male to mate with her in exchange for being permitted to carry on living in what is now his territory. The female is intent on living in a particular territory and is prepared to mate with any male who successfully lays claim to that territory in order to do so. In principle, this is still prostitution – the trading of sex for resources – even though it is taking place within a monogamous relationship. As such it is little different from the behaviour of the majority of women around the world, few of whom would consider themselves to be prostitutes.

Men, also, of course can prostitute themselves. The young gardener in Scene 18 is an example. In most circumstances, however, it is much more difficult for a man to find a woman who is prepared to pay, in whatever sense, for sexual access to his body. Most men are only too willing to have sex with a woman without any reward other than the sexual opportunity itself. In contrast, for all the reasons we have discussed in this book (see particularly Scene 28), women have much more to lose from any single sexual encounter than men and usually need some sort of pay-off to balance the potential cost. Only when a woman is eager to collect a particular man's genes might she be prepared to pay in some way for the privilege.

However we define prostitution, though, the woman in Scene 32 is clearly at the 'empid fly' end of the scale we have just described rather than the 'bird' end. For her, prostitution is a way of life. Biologically, it is also a strategy for reproduction – and, moreover, a highly successful one. By her mid-thirties she had had four children and had earned more than enough money to give them a comfortable and healthy environment in which to live. Each child had a different father, at least two of whom could have been men of status. What all of the fathers had in common was that they produced very competitive ejaculates: sperm armies able to defeat those of many other men. The woman's sons, her grandsons, and any later male descendants should also have an above-average chance of producing competitive ejaculates. In later generations, many people in the population would inherit her genes because of the competitive success of her male descendants.

This advantage of prostitution as a reproductive strategy is the same advantage gained by any woman who promotes whole-sale sperm warfare in her tract (Scene 21). It is just that prostitutes exploit this technique more often than any other category of women in the population. Few other women, except those who suffer gang rape (Scene 34) or who seek out group sex (Scene 21), are ever likely to contain the sperm from as many different men at the same time.

The success of prostitution as a reproductive strategy means that most of us will contain the genes of a prostitute among our ancestors. On average, we should each need to go back through our family tree no further than the 1820s (seven generations) before finding an ancestor who was born to a prostitute (assuming, conservatively, that only 1 per cent of the population is conceived by overt prostitutes).

As a way of life, however, prostitution has many risks. First and foremost, as we have discussed on numerous occasions, there is a high risk of contracting sexually transmitted disease. This factor alone can condemn a prostitute to early infertility and death. Many try to reduce the danger by using condoms but are forever fighting against male aversion (presumably for the reasons discussed in Scene 29). Prostitutes report men trying surreptitiously to remove the condom that they have agreed to wear. Even since the advent of AIDS, the majority of clients prefer not to use a condom. So strong is the male body's preference for insemination that, like the woman in Scene 32, many prostitutes are prepared to settle for simply exploiting the situation by charging more for intercourse without a condom.

There is also the danger of being injured or killed by a client. Prostitutes attempt to reduce this risk by gathering together in a brothel or massage parlour, by paying a man or men to watch over them and provide protection, or simply by having a chaperon (often their mother or father).

By far the biggest danger to prostitutes, however, is drug addiction – a danger both physically and financially. Despite sometimes having to pay rent on the room from which they work, or for protection, prostitutes have an earning capacity that most other people can only fantasise about. Yet few prostitutes are rich. In part, this is as a result of exploitation: young girls are first introduced to the drug scene and then 'encouraged' into prostitution as the only means of earning enough money to support their habit. The agents then take their own commission for providing 'protection'. Such sad cases inevitably incur the costs of prostitution without reaping the benefits.

Not only is prostitution a reproductive strategy for women, the visiting of prostitutes is also a reproductive strategy for men. In ancient Greece and Rome, almost all men inseminated a prostitute at some time in their lives. In the USA in the 1940s, 69 per cent had inseminated a prostitute at least once and 15 per cent did so on a regular basis. In the UK in the 1990s, 10 per cent of men between forty-five and fifty-nine years old have paid for sex at least once in their lives. Apart from the occasional priest, the average man who pays for sex usually has other reproductive outlets – in the UK, for example, they are also more likely to have an above-average number of unpaid partners.

As a reproductive option for men, prostitutes provide a target for insemination that is quick and easy. The more prostitutes a man inseminates, the greater his *potential* reproductive success. He may also gain in other ways, depending on his situation. For example, the young student in Scene 32 was hoping to gain experience (Scene 27) that would later be of use in his more conventional reproductive attempts with other women (on average, students lag two years behind other people in their sexual experiences). Men without a partner are simply including prostitutes in their search for females to inseminate. Occasionally, a man may find a long-term partner through his contact with prostitutes, like the woman's partner in Scene 32. Men already with a partner are using prostitutes as targets for their infidelity.

Prostitutes are also potentially expensive, both financially and in terms of disease risk. Thus, although insemination is easy, as we have just seen, the reproductive benefits are relatively low.

Any single ejaculate has a low chance of achieving fertilisation because of the intensive sperm warfare that it will encounter. In addition, even during long periods of unprotected sex, prostitutes are less likely to conceive in any given month than a woman who is having only occasional sex with a single partner. This could be due to prostitutes having a lower rate of ovulation (Scene 16) or a more efficient cervical filter (Scene 22). It could also be due to sperm warfare. Perhaps such warfare is so intensive inside prostitutes that the sperm armies from different men often neutralise each other.

A man needs to inseminate prostitutes many, many more times than he does a conventional mistress in order to produce a single child outside of his long-term partnership. Of course, no man would say that his reason for visiting a prostitute is to gain a chance of fathering a child. Nevertheless, from time to time a man will succeed in this, passing on to his child in the process the genetic program that made his ejaculate so successful at sperm warfare – prostitutes are unsuitable targets for men who are not sperm war specialists (Scenes 19, 30, and 35).

The final conundrum is why some men are prepared to form long-term partnerships with prostitutes. Clearly, such a man experiences all of the disadvantages just discussed for a prostitute's clients. Usually, however, he trades these costs against the benefit of a share in her wealth. The man in Scene 32 traded his financial and lifestyle benefit for effectively looking after her, her home and her children. In addition, he had some opportunity to father a child with her and may already have done so. His lifestyle probably also meant that he would have a chance to inseminate other women as well. His reproductive strategy was risky. The potential gains were high, but so too were the potential costs. Apart from any other genetic characteristic he needed to benefit from such a strategy, at the very least it would have helped if he had also been a sperm war specialist (Scene 35).

Given that prostitution and the use of prostitutes are both

reproductive strategies with high potential benefits and considerable potential risks, prostitutes share many similarities with bisexuals (Scenes 30 and 31). We do not know whether there is a genetic predisposition for prostitution in the same way as there is for bisexuality, but if there is, it would seem that only a minority of women possess the genes for prostitution. So it is possible that, as we concluded for bisexuality, prostitution is an advantageous strategy only as long as it remains relatively uncommon, at least in societies in which its costs are high. If all females were available to all men, the potential value of prostitution and the reproductive edge it would give each prostitute over other females would disappear. At the same time, the spread of disease would greatly increase.

If this 'genetic minority' interpretation of prostitution is correct, the corollary, as for bisexuality, is that on average the reproductive success of prostitutes and other women should be *the same*. Without a genetic analysis similar to that which has been carried out for bisexuals, however, we cannot be certain – and there is an alternative interpretation. This is that all women are *potential* prostitutes – but that only a few ever encounter a situation in which they judge the potential benefits to outweigh the potential costs. If this interpretation is correct, the corollary is that women who judge the situation correctly and become prostitutes should have a *higher* level of reproductive success than other women.

This interpretation has less in common with our interpretation of bisexuality (Scenes 30 and 31) than it has with our interpretation of another high-risk, minority sexual strategy – rape. This is the subject of the next two scenes.

SCENE 33

The Predator

The man locked his car, then made his way along the dark road. As he walked, he could hear the sound of traffic a block away on the main road running through the town. It was midnight but still hot, and the pavement cafés and bars around the busy main square were still crowded and noisy, mainly with holidaymakers. Turning away from the square, he saw at the end of a narrow road the beacon he was seeking. It was a telephone booth, shining out of the darkness of a marble-tiled square outside a church.

He phoned home, checked with his partner that she and their two children were well, and told her he would see her the following night. With that, he walked a short way back down the road, stepped into the darkest of shadows, and leaned back against the cool wall of the church. After checking the flick-knife in his back pocket, he lit up a cigarette and settled back to wait. Apart from the glow of the cigarette, he was totally invisible. His throat dried and tightened in a mixture of excitement, fear and anticipation. The waiting was the part he liked best.

The girl at the pavement bar was crying. The anger had gone but the pain and worry were still there. Two tables away, laughing drunkenly with his friends, was this holiday's sexual partner. As she sat with her friends, she could feel the flowback dampening her knickers. Scarcely more than an hour ago, they had had sex on the grass in the nearby park.

She wasn't on the pill. All week, since they first met, he had used a condom. Tonight, he had only pretended to put one on. As if that hadn't been cruel enough, no sooner had he withdrawn than he had gloated that she was the second that night – that an hour earlier he had been in the same spot with another of the girls from their group. He always liked to have more than one girl each holiday, he boasted, and if she really wanted to know, he much preferred the other girl, anyway. But he didn't mind carrying on having sex with her for the rest of the holiday, if she wanted.

After insulting him to the limits of her vocabulary, she had made her way drunkenly back to the bar to seek the sympathy of her friends. Now, after the anger and the tears, she had the sudden urge to speak to her boyfriend at home and to seek the reassurance that at least he still cared for her. Clumsily, she stood up, her chair falling over on to the pavement. She told her friends she was going to the phone. Her best friend offered to come with her, but she said she wanted to be on her own.

The man had been leaning against the wall in the shadows for about fifteen minutes and was smoking his second cigarette when he saw the girl appear at the end of the road. In all that time, only one couple had come to the phone and the church square was now empty. He waited, fully expecting someone to be with the girl. But no, she was on her own. His luck was in, yet again. For a few moments, she was back-lit by the lights from the main square and he could see the silhouette of hips and legs through her thin dress. Then she was in shadow and he heard, rather than saw, her coming towards him. He took one last drag on his cigarette and threw it away. His penis sensed action – beginning to stiffen into the rod of iron that hadn't once let him down. This girl would be his fifth this summer. He reached into the back pocket of his trousers for the knife.

The girl had noticed the glow of a cigarette in the shadows but was too drunk and too angry to sense danger. She barely noticed the man as she passed him in the blackness, but a second later she was grabbed from behind, a strong arm around her throat. The knife in front of her face was silhouetted against the light of the distant telephone kiosk. As he held her, she glimpsed a group of people walking past the end of one of the other roads leading away from the church. She wanted to scream, but fear of the knife, the man's arm around her throat, and blind panic left her totally mute.

From that first moment, everything seemed to happen at the speed of light. Still standing behind her, the man pressed the cold metal against her cheek and ear. Speaking softly, in a language she didn't understand, he moved his arm from her throat to her waist and made her bend forward, as if to touch her toes. He lifted her dress forward over her head, placed his knife in her knickers and, with two quick, practised cuts, removed them. Within seconds of making her spread her legs, he was inside her. Fifty thrusts later, he ejaculated. For a few seconds, he paused before withdrawing and for one panic-stricken moment she thought he was going to kill her. But he pushed her to the ground and ran off into the darkness. Within a minute he was back in his car, driving to his hotel, two towns away.

The girl lay sprawled on the ground in shock for several minutes. Then came the tears. Eventually, she got up and moved off, a sad, dishevelled figure. She saw no one. Back in her hot, humid hotel room, she took off her clothes, stepped into the shower, and stayed there for nearly an hour, washing and washing herself. She fell on to the bed and spent the night half in shock, half in sleep.

Inside her body, a sperm war raged between the armies from her holiday lover and the rapist, while her right ovary made preparations for ovulation. Two days later, while she was still agonising over whether to tell her friends or the police, she conceived. The war was over.

Daunted by the prospect of explaining what had happened to police who did not speak her language, and worried by how her boyfriend might react if he ever found out, she told no one about the rape. A week later, back in her own country with her boyfriend, she contrived on their first night together to have unprotected sex.

On discovering her pregnancy, she first considered an abortion, but with her boyfriend's reassurances that he would help raise what he thought was their child, she decided to keep it. By the time her baby was born, she had managed to convince herself that it was probably his anyway. But it wasn't. Her son's father was in fact a predatory rapist from a distant country.

~

Biologists are rarely less popular than when they offer an objective analysis of rape. If they say that rape yields reproductive benefits, they are accused of encouraging the act. If they conclude that rape has a biological basis, they are accused of condoning it. If they dare to suggest that female behaviour may sometimes invite rape, they are accused of violating womankind just as surely as if they had committed the act themselves. Yet to report proper conclusions is not automatically to encourage, condone or violate. Are historians accused of encouraging warfare if they conclude that a country benefited from a war? Are they accused of condoning warfare if they conclude the behaviour has a biological basis? Are they accused of violation if they conclude that a country invited invasion? Or are they, on the other hand, congratulated for incisive historical analysis and for aiding the prevention of future conflicts?

Prevention of socially unacceptable behaviour, be it rape or warfare, demands a proper understanding of the circumstances that trigger such behaviour. If the objective conclusion is that under certain circumstances all men will rape or all countries will go to war, there is no value in wishing or pretending that men or nation states are somehow different. The value will lie instead in identifying the circumstances that facilitate rape or war. Then, some attempt can be made to ensure that those circumstances rarely, if ever, occur. The only reliable pathway to understanding the situation is via objective analysis, even if en route some conclusions may be unpleasant, unpalatable or socially 'incorrect'.

The first step is to be clear about the phenomenon that we are about to discuss. The rape in this scene was a sexual event which had a reproductive outcome. The girl was violated, but not physically damaged, and was left able to conceive and give birth. The rapist's weapon – in this case a flick-knife – was used as a means of coercion, not to inflict physical damage. This is the commonest scenario. As testament to the *reproductive* nature of their behaviour, such rapists usually select victims who are at their peak of fertility, aged between twenty and thirty-five.

On such occasions, the bodies of the men concerned are clearly using rape as a reproductive strategy. On a minority of occasions, however – albeit those which receive the widest publicity – a woman is not only raped but is also physically harmed, mutilated or killed. On these occasions the men concerned are bent on violence and murder, not reproduction. As testament to the *non-reproductive* nature of their behaviour, such rapists are more likely to select older women, from their late thirties onwards – women who are past their peak of fertility.

These latter, more violent events are not part of our discussion. Here we are concerned solely with sexual events in which the woman is not physically harmed by the rapist, at least not so much that she is unable to conceive from the intercourse or later give birth (though she may suffer psychological consequences). On these, the majority of occasions, the reproductive success of both the rapist and the victim is influenced by their interaction. Moreover, such rape often involves sperm warfare. This relationship between rape, sperm warfare and reproductive success is an essential and proper topic for consideration in this book.

We have to ask whether rape can be a successful alternative strategy in the male pursuit of reproductive success. And, unpalatable as the question may be, we also have to ask whether conceiving via a rapist attack can be an alternative strategy for the female body, as well. Anthropologically, rape of the type described in Scene 33 is *common* in nearly 50 per cent of societies and is reportedly *rare* in only 20 per cent. It is estimated that in some of the world's major industrial cities nearly half of women experience an attempted rape in their lifetime, and a quarter are actually raped. Such estimates are, of course, very approximate and confounded not least by the fact that, as in our scene, very few rapes are actually reported. The current estimate is that only about one rape in ten is made known to the authorities. So widespread is rape and so often do children result from rape that all of us could probably find a rapist among the past five generations of our ancestors.

Humans are by no means the only animal in which rape is part of the male sexual repertoire. The males of species as diverse as insects, ducks and monkeys are all known to force copulation under certain circumstances.

At least one insect, the male scorpion fly, has a special hook on its wing which it uses to grip the female while it forces copulation. Without that hook – for instance, if it is removed by an experimental biologist – the female always escapes a male's attempt at force. All males in this species have a hook, but not all males need to use it. Females are attracted to the bigger males and allow them to mate without being forced. Rapists in this species are the smallest males, unattractive to females, who cannot obtain a mate in any other way. Even so, they are relatively unsuccessful and are lucky to force a mating more than once in their whole lives. In contrast, the largest males have females queuing up for their unforced favours. In this insect, therefore, rapists are males making the best of a bad job, having been born small and relatively unattractive to females. This is not the case for most animals.

More often than not, rape is an extra option for males who are in all other respects reproductively normal and attractive. Among white-throated bee-eaters (a species of bird), for example, the rapists are individuals who, having managed to breed early in the season and having raised young with a monogamous partner, later embark on raping forays. They chase any female who is still fertile and who has been left unguarded by her mate, and attempt to force insemination. Sometimes such females offer resistance. At other times, like the girl in Scene 33, they may simply allow the male to mate. However the female birds behave, the outcome is the same. Not only do rapists produce young with their partner, they also produce a few through forcing copulation on other females. The result is that, on average, rapists of this species have a greater reproductive success than non-rapists.

As we shall soon see, it is crucial for our understanding of human rape, particularly the response of the female, to know whether rapists have an above- or below-average reproductive success. Are human rapists more like the scorpion fly, or the white-throated bee-eater? Are they reproductively below average – making the best of a bad job after a lack of success via more conventional strategies? Or are they above average – adding their success via rape to an average success achieved more conventionally?

Unfortunately, the evidence is a little ambiguous. There have been a few, albeit rather weak, claims that rapists are, in fact, like the scorpion fly, making the best of a bad job. Such studies describe the average rapist as young, poor, and physically unattractive to women. Other much more thorough studies have shown that age for age, social status for social status, and physical attractiveness for physical attractiveness, rapists are no less likely than non-rapists to have a partner and children. If we believe these latter studies, it would seem that on balance human rapists, like white-throated bee-eater rapists, have the potential for an above-average reproductive success. Whether they realise that potential, however, depends on whether they succeed in making the potential reproductive advantages outweigh the potential costs. Like the other minority strategies discussed in this section (bisexuality, prostitution), rape is a risky business. Not only is there the ever-present possibility of infection, there is also a real danger that rape will trigger strong and often violent retribution.

First, there is some risk that rapists will be injured or even killed by the victim herself – though, as in the scene, females do not always resist. Secondly and more importantly, they are at risk to retribution both from their victim's partner and from the society in general. Men do not stand idly by and watch their or other men's partners being raped if they can prevent the attack. Vigilante groups against rapists are by no means uncommon, particularly in large cities. Nor are humans alone in showing such defensive behaviour. The male members of a lion pride, for example, will band together to repel attacks from marauding gangs of bachelor males intent on gaining sexual access to the females. Injuries, sometimes fatal, are commonplace.

In most modern human cultures, however, most of the defence against rapists is institutionalised. Incarceration rather than violence, injury or early death is a rapist's destiny if he is ever caught.

Rape is clearly a dangerous reproductive strategy. It can be very successful but it can also be a complete failure. As a result, the successful rapists amongst our ancestors, those who left most children and thus whose genes we are now most likely to possess, were those who were the best judges of circumstances; those who most accurately assessed (subconsciously) whether the potential reproductive benefits of an opportunity to rape really outweighed the potential costs.

So far, we have looked at rape as a reproductive strategy for men. When we consider women and their physiological response to being raped, we encounter an unpalatable conundrum. Theoretically, if being raped is reproductively disadvantageous to a woman, she should be *less* likely to conceive from it than from routine intercourse. On the other hand, if being raped is advantageous, she should be *more* likely to conceive from it than from routine intercourse. So what happens? Are women more likely, or less likely, to conceive via a rapist?

All of the available evidence points in the same direction. It suggests that, like the girl in Scene 33, a woman is *more* likely to conceive from rape than from routine sex with her partner. It is just possible that this conclusion is false, the data reflecting instead a greater probability that a woman will report rape if she is likely to conceive. But this is unlikely for two reasons. First, a woman has to report a rape long before she knows whether she has conceived. Secondly, the difference in conception rate between rape and routine sex is greatest during the least fertile phases of the menstrual cycle – during menstruation and again three or more weeks after menstruation. These are precisely the times that a woman would least expect to conceive.

However reluctantly, therefore, we should accept that a woman is more likely to conceive from rape than from routine sex. The most plausible explanation is that the trauma of rape actually stimulates ovulation, especially if her body happens to be 'on hold' (Scene 15). We have already seen that rather violent rough-and-tumble sex play can trigger ovulation in mink (Scene 28).

However, trauma may not be the only factor. Although a woman is more likely to conceive from rape than she is from routine intercourse with her partner, there are other, less traumatic, situations in which she is equally likely to conceive. One is when she has sex with a partner whom she has not seen for a long time and whom she sees only briefly – as when a soldier comes home on a short leave (Scene 15). Another is during snatched moments of infidelity (Scenes 6, 17 and 19). What these two situations have in common with rape is not trauma but a limited opportunity to collect a particular man's genes. The physiological mechanism for conception is also likely to be the same in all three situations – ovulation in response to intercourse (Scene 15).

It is not difficult to see why a woman should want to make the most of a limited opportunity to collect the genes of a partner or a lover. But why should she want to make the most of a one-off opportunity to collect the genes of a rapist?

It was to answer this question that we needed to consider earlier whether human rapists have an above- or below-average reproductive success. As we have seen on many occasions in this book, the genes of males who have an above-average reproductive success are desirable targets for a woman's body. If she can collect such genes, she will increase her reproductive success via male descendants who inherit the same potential for success. Since we decided for rapists that, on balance, they do indeed have above-average potential, it should be no surprise to find that when a woman's body has a one-off opportunity to collect a rapist's genes, it often does so.

This conclusion does not mean, as people often assume, that a woman should therefore seek to be raped. On the contrary, it is reproductively important to the woman that her body collect genes from only the most successful of rapists. If she conceives to an inept rapist, doomed quickly to be caught and to suffer social retribution and incarceration, her male descendants would inherit unsuccessful characteristics. As we discussed earlier (Scene 28), a man needs to pass certain tests if his genes are to be acceptable to a woman. The only way a woman's body can select out the most successful of rapists is to do everything possible to avoid being raped. She should avoid risky situations and take full advantage of the protection offered by her partner, by other individuals, and by the wider society. Whether she should also try to fend off the rapist physically will depend on her assessment of the danger of being physically damaged. Unlike the tests set by women during rough-and-tumble sex play (Scene 28), she may often be better advised to follow the example of the girl in Scene 33 and not resist. A woman who follows this overall strategy is unlikely to fall victim to any but the most cunning, determined and competent of rapists. The result is that only a minority of women are ever raped, but those who are may then respond by conceiving.

Our discussion should not end here. If rape can be a successful alternative reproductive strategy for men, and if conceiving via the more competent of rapists can be a successful reproductive strategy for women, we need to ask why rape is not more common. In particular, we need to discuss whether rapists are a genetic minority, like bisexuals (Scenes 30 and 31), or whether all men are potential rapists. This question will be examined in Scene 34.

The only remaining question to answer here is why the rapist in Scene 33 was the winner of the sperm war inside his victim's body. In fact, the contest was very one-sided. First, the girl's boyfriend, because he normally used a condom, failed to enter an army until it was too late. The fertilisation trophy had already been claimed. Secondly, all week her holiday partner had also been using a condom. So when the day of our scene began, the girl contained no sperm. On the night of the rape, her holiday lover actually inseminated her. But this made little difference because his body, unable to allow for having used a condom previously (Scene 29), will have calculated that few sperm were needed to top her up, it being only a day since they last had sex. In addition, he had just inseminated another female and, as it was their first intercourse, he would have given her a large inseminate. The result was that the victim received from her holiday lover only a small inseminate, one rich in young killers and egg-getters but relatively low in immediate blockers. In addition, she had ejected the flowback from that intercourse before being raped.

The overall result was that, when the rapist inseminated the girl, his seminal pool had her vagina to itself. Moreover, his sperm were confronted with only a minimal cervical filter and a small army of killers. He will have injected a large army, full of killers and egg-getters. Admittedly, the holiday lover had an hour or so's start, but ovulation was still two days away. It is likely that by the time the egg-getters in her oviduct encountered an egg, they would virtually all have been from the rapist. Another set of rapist genes would have been passed on to the next generation.

SCENE 34

Soldier, Soldier

A single shot rang out. A flock of large black birds flew noisily out of the trees as the five soldiers threw themselves to the ground. While the men lay quietly in the cover of vegetation by the side of the road, the birds circled briefly above them before re-settling one by one in the trees. Quietness returned. The soldiers looked at each other, checking first whether anybody had been hit, then whether anybody had any idea where the shot had come from. They waited for a while, then crawled along the roadside on their knees and elbows, seeking the cover of trees just ahead of them.

Once sheltered, they discussed the situation. A few minutes ago, one of them had noticed a house. Almost hidden by trees, it was by the side of a rough path that joined the road. A sniper in the house might just have had one chance for a shot at them through the trees as they passed. They decided to investigate and planned their approach for maximum cover, splitting into two groups: three to enter the front of the house, two to go in by the back.

The house was small and in poor repair, but obviously inhabited. There was no sign of a potential sniper at the windows, but the soldiers took no chances in their final approach. It was very hot and the group of three were soaked in the sweat of exertion and tension as they paused by the front door. When they burst in they found nothing more threatening than an elderly couple, a baby, and a young girl of about twelve. Nevertheless, the soldiers remained tense and alert. The girl ran screaming to her grandmother, who was holding the baby. Adrenaline still coursing through his veins, the soldier in command asked the grandfather who else was in the house. 'Nobody,' said the old man, fear in his eyes and voice. The soldier hesitated, very much on the edge. He pointed at the children and asked the whereabouts of their parents. The old man looked at his partner, then said they were dead, shot by people like them.

The soldier looked around, told the old man that if he was lying he also was dead, then ordered his two companions to search the house. The pair moved tentatively towards the only other exit from the room. It had no door, but a makeshift curtain. As they reached the doorway, there was a sudden movement in the room beyond. They dodged quickly to either side of the exit, guns raised. One of them reached for the curtain just as they heard the back door burst open. The other two members of their patrol had arrived. There was the sound of scuffling, then the curtain was thrown back and the two soldiers appeared, pushing a woman in her late twenties in front of them. She fell to the floor. As she raised herself to her knees, her daughter ran to her and clung round her neck, sobbing.

The soldier in charge pointed his gun at the old man.

'You lied, you're dead,' he said.

'You *did* kill their father,' the old man replied.

'And now we're going to kill you,' continued the soldier, 'but maybe not yet. I think you deserve to watch us have our fun, first.'

He ordered the others to check the house, inside and out. While they were out, he stood over his captives, gun in hand, waiting and smiling. The others returned. He ordered one to watch the back, another the front, and the other two to watch the old couple. The soldiers knew what was coming, and began to relax. This wasn't the first time during this war they had been in this position.

Their commander looked at the mother and her twelve-year-old daughter, sobbing in each other's arms. Then he shouted to his companions, asking them which it was to be, the mother or the daughter. The mother shouted hysterically, clutched her daughter, and begged the soldier to take her, not the child. The soldier pulled the girl away from her mother and threw her back towards her grandmother. Then, prodding the mother with his gun, told her to stand up and take off her clothes. When she was naked he pulled her over to a table in the corner of the room, making her bend over to rest her face and chest on the table. He had chosen the location carefully, so that while standing behind her he could still see the room. Once satisfied with his vantage point, he threw his gun to one of his companions, opened his trousers, and entered the woman from behind. Throughout the intercourse, he was coolly vigilant, only losing his concentration during ejaculation itself. As he removed himself, he told the woman to stay where she was. Then he walked over to retrieve his gun and told the soldier who had been holding it that he was next.

One by one the soldiers took it in turns to inseminate the woman in front of her daughter and parents. Throughout her forty-minute ordeal, the woman occasionally sobbed but, afraid that they might turn on her daughter instead, did not once ask them to stop. When all of the soldiers had had their turn, she slumped to the floor in the corner, unable to look at anyone. The soldiers allowed her daughter to run over and join her, but made her stay naked, not allowing the grandmother to take over her clothes.

For a while, the soldiers sat around, smoking and laughing, recovering from their exertion. Inevitably, somebody eventually suggested that maybe they should have the daughter anyway. A heated discussion broke out. Two of them voiced their disapproval and swore at their companions for even considering the prospect. But the soldier in command was keen. He told the other two they could abstain if they wanted, but he was going to make the most of their find. When the mother tried to stop him from taking her daughter, he knocked her unconscious with the butt of his gun.

When three of them had raped the young girl, their commander decided it was time to leave. They were just about to open the front door when he told his companions to stop. Turning to the grandfather, he asked the old man if he had really thought he would forget to do what, on first arriving, he had promised; then he shot him. Next, he turned his gun towards the old woman, who was still holding the baby. He paused, finger on trigger, then smiled. Telling her that her family had given him enough fun for one day, he lowered his gun.

Two children were conceived that day, one by the mother and one, against all odds, by the twelve-year-old girl. However, the soldiers would never know nor care which of them was the father, in either case. No sooner were they clear of the house than they were all dead, gunned down by guerrillas. The young girl's father had seen the soldiers approaching and gone for help.

~

We have already discussed the presence of predatory rape in the male sexual repertoire (Scene 33). Our conclusion was that rape is a strategy by which a man can increase the number of women who might have his children. Any child sired in this way is a bonus, an addition to those fathered via a conventional long-term relationship. And in order to gain this bonus, a rapist has to negotiate successfully the defences that surround a woman. Rape is a risky strategy, and incompetent rapists lose out rather than gain from their behaviour.

We have also discussed why women should do everything possible to avoid being raped. Having been raped, however, a woman may gain, reproductively, by then conceiving.

All of the arguments rehearsed in Scene 33 apply with equal force to the 'gang rape' we have just witnessed (which is loosely based on a documented event in a recent war). Before examining gang rape in humans, we shall first look at the phenomenon in other animals. Males of at least one species of monkey are known to band together for this purpose. But the most detailed information is available for birds, in which gang rape is particularly common and has been described for many species. On his own, a male bird finds it almost impossible to force copulation on a female. Intercourse in birds involves the male balancing precariously on the female's back, then bending his tail under hers. Most do not have genitalia as such, but simply a small sac which they have to evert. Male and female sacs are similar, and sperm are transferred by the male and female pressing their sacs together once the male is in position. On his own, a male can easily be dislodged from a female's back. Moreover, she can avoid accepting his sperm simply by not everting her genital sac. A group of males, however, can effect copulation by forcing her to the ground through weight of numbers and then continuing to peck and attack her until she accepts sperm transfer. Females who do not comply have been known to die in the mêlée.

There are more benefits to gang membership than simply being able to force insemination. A gang is also more able to brush aside the female's defences, either her own or those of guarding males. In some species of duck, for example, the gangs divide their labour. On discovering a female and her partner, some of the gang chase and keep away the guarding male, while the others rape the female. Members of the gang swap roles, so that over a period all benefit from their membership.

The *benefit* of gang rather than individual rape to male birds, therefore, is that through cooperation they each increase their chances of inseminating an unwilling female. The *cost* is that on each occasion their sperm have to engage in warfare with sperm from other males in the gang. Whether a male gains an overall advantage from being a gang member depends on how these costs and benefits balance out. Mainly, it depends on whether the greater number of females a male inseminates through gang membership compensates for his reduced chance of fertilisation on each occasion. If he is a member of a gang of four, he has only a one in four chance of fathering any child produced by their activities, so only if the gang as a whole manages to inseminate more than four times as many females as the male could on his own is his membership worthwhile. For this reason, gangs tend to be rather small. A group of ten will not be twice as effective at raping females as a group of five – yet a male in the larger gang has only half the chance of winning sperm warfare of a male in the smaller.

As far as humans are concerned, a significant proportion of all predatory rapes are gang rapes – even in peacetime. Some estimates suggest that 70 per cent in industrialised societies are gang rapes. Others suggest about 25 per cent. Inevitably, human males gain all the same reproductive benefits from gang membership as other male animals. The same arguments apply and the size of human gangs tends to be small, usually four or five individuals.

In our discussion of predatory rape in Scene 33, we left open one very important question. If rape can be a successful alternative reproductive strategy for men, and if conceiving via the more competent of rapists can be a successful reproductive strategy for women, why is rape not more common? There are two extreme possibilities, just as there were for prostitutes (Scene 32). One is that rapists are a genetic minority, like bisexuals (Scenes 30 and 31). The other is that all men are potential rapists – but that rape still remains rare because few men ever encounter a situation in which the potential reproductive benefits of rape outweigh the potential reproductive costs (many of which are imposed by the wider society (Scene 33)). Of these two possible explanations, the latter seems likely to be nearer the truth – because in wartime the incidence of all forms of rape, including gang rape, increases dramatically.

There are three main reasons why this is so. First, having

routed its enemy, a successful army has effectively removed the males who might otherwise defend the women. Secondly, the mobility and confusion of war make it very difficult for rapists to be tracked down. Thirdly, because of the ever-present risk of imminent death, the prospect of social recrimination for rape seems trivial. On balance, therefore, unlike in peacetime, men in the midst of war are much more likely to chance across a situation in which the reproductive benefits of rape outweigh the costs.

The increased incidence of rape during warfare is due to more men becoming rapists, not to a limited number of men raping more often. This pattern is much more what we should expect if all men were potential rapists, rather than if only a fixed genetic minority had rapist inclinations. Unpalatable though this might be, it should be no more so than that other facet of male behaviour highlighted by war – that all men are potential killers.

The soldiers in Scene 34 were essentially ordinary young men who, had they been born at another time in another place, would have proclaimed themselves incapable of either rape or killing. Yet, in a wartime scenario, they found it within themselves to commit both. This should not be so surprising. As we discovered in Scene 33, we all possess the genes of rapist ancestors. For the same reasons, we all possess the genes of past killers.

First, let us consider wartime killing. Few people would deny that the predisposition to wage war against neighbouring communities has a biological basis. Similar behaviour is shown by many social animals, from insects to primates. Neighbouring communities of chimpanzees, for example, each forty or so strong, have been seen to engage in what can only be described as inter-group warfare. One social group was observed over a period of several months systematically murdering individuals from the other group, until none were left. As far as humans are concerned, historical inter-village battles in New Guinea and South America should be regarded as little different, either in scale or in territorial motivation. Only the distance from which the people could kill each other was different. And although modern human warfare between nation states is clearly greater in terms of scale and mayhem, the behaviour, motivations and fears of the individuals concerned as they kill each other are just the same.

Distasteful though it might be to accept, most people are alive today only because at some time in the past the wider society in which their ancestors lived successfully waged war and killed people. That society either successfully attacked or defended itself in warfare against an enemy intent on territorial defence or expansion. History books are full of civilisations and societies that have been annihilated by their neighbours. Obviously, we are more likely to be the descendants of the killers, and not of the ones who were killed, so we all have the potential to kill. Whereas in wartime, many people are continually encountering circumstances in which the benefits of killing another person far outweigh the costs, in peacetime far fewer people encounter such circumstances.

Next, let us consider wartime rape. We have all inherited our genes from those men amongst our wartime ancestors who took their opportunities to reproduce when they could, even when it involved rape. We have not inherited the genes of men who opted to wait for a safer and more conventional opportunity in a future that for them never came. Those men in Scene 34 who decided *not* to rape the young girl did not live to produce more children who would inherit their compassion, whereas one of the men who raped her did produce a child to inherit his lack of compassion. It is by this process of weeding out genes that do not enhance reproductive success that evolution has saddled the majority of men with the propensity to behave as rapists in the appropriate situation.

Wartime rape and murder also combine to shape our characteristics in the process known as *genetic infiltration*. Warfare displaces and sometimes annihilates whole lineages by this process. The invading armies annex lands and rape the women. Then subsequent settlers establish relationships and have children in less aggressive ways. By this means, whole genetic stocks can be infiltrated and diluted almost out of existence by more successful genetic lines. There are few people in the world today who do not contain at least a few genes from ancestors who, two thousand years ago, lived in a completely different geographical area. Wartime rape is one of the key elements in this ebb and flow of genes across the earth's surface. Not one of us would be precisely the person we are today if one of our ancestors had not taken advantage of the relative immunity of war to commit rape and murder.

Biologically, it might seem strange that the men in Scene 34 considered a twelve-year-old girl a suitable target. But, unfortunately, it is not. Authenticated records show that women have conceived at all ages between seven and fifty-seven, with unconfirmed reports of women conceiving up to the age of seventy. Reproductively, therefore, a predisposition to find any woman between these ages to be sexually attractive is not totally futile. Pre-pubescent girls may ovulate before their breasts and pubic hair develop and before their first show of menstrual blood. Post-menopausal women may conceive up to at least eighteen months after their last period. In between, as we have discussed at length (as in Scene 2), men are totally unable to judge whether and when a woman is fertile.

As far as rape is concerned, young girls and older women are in a sense victims of their gender's success at unconsciously confusing and deceiving men. Unable to decipher female fertility, the male body has responded evolutionarily by a blanket approach – inseminating whoever and whenever it can. As we have noted, given the choice it will prefer women in their most fertile years (twenty to thirty-five), but under some circumstances it will find even very young and very old women attractive. Faced with four females, one a baby, one twelve years old, one in her late twenties and one in her sixties, all of the men in Scene 34 wanted to inseminate the woman in her late twenties, three wanted to inseminate the twelve-year-old, and none wanted to inseminate the older woman or the baby. Under different circumstances, with a more restricted choice for the men, the older woman might also have suffered rape.

The two who were raped both ovulated and conceived. Perhaps they were stimulated to ovulate by the rape itself, as we discussed previously (Scene 33). This was almost predictable for the mother, but was against the odds for the daughter. Given that females *might* ovulate in response to rape, the chances of the mother conceiving could have been as high as about one in three. The chances for the daughter conceiving, however, were probably no higher than one in fifty.

It is difficult to guess which of the soldiers sired the two children. Inevitably, one of the features of gang rape is that it generates intensive sperm warfare in the victim. All else being equal, we might expect the first male who inseminated each female to be the reproductive winner (Scene 21). He was the one whose egg-getters got a head start in reaching the oviducts, whose killers were first to take up their positions in the womb, and whose blockers had the chance to establish themselves in the cervix. However, as we discussed in Scene 21, much will have depended on how quickly the subsequent men also inseminated the female, on how efficiently each penis removed the seminal pool from the man before, and on precisely when each female ovulated. In Scene 34 the odds will have favoured the soldier in command because he was the first to inseminate both females, but any one of the doomed men could have left a descendant to perpetuate his rapist genes.

SCENE 35

Men Are All the Same

The sound of laughter erupted from the corner of the bar. The two women at the table looked around, suddenly feeling conspicuous – but they had drunk enough not really to care. They leaned towards each other again, the taller of the two making the shape of a cup with her free hand, determined to convey to her friend just how big her partner's testes were. She'd seen quite a few men's packages in her time, she said, and her partner's were by far the biggest.

The two hadn't met for years but had exchanged the occasional letter. Finally, they had decided it was time to meet up again, so the other woman had invited her friend, her partner and their two young children to stay for the weekend. Tonight, the men had wanted to watch some sport on the television, so the women had decided to leave them to baby-sit while they went out to get drunk and catch up on each other's lives. Now that they *were* drunk, what they really wanted to talk about was sex. In particular, they wanted to know about each other's sex life.

The other woman wasn't sure that she would like big testes. Her own partner, she said, trading information, had really small ones, smaller than table-tennis balls. His penis seemed a bit small, too. Not wanting to miss the opportunity, the taller woman said her partner's penis was definitely big. Sometimes she wished it weren't quite so big. He tended to be really vigorous when he was thrusting and sometimes he hurt her. She didn't know what it was inside her that he kept poking, but it was certainly uncomfortable. She also wished he didn't want sex quite so often. Most of her friends, she said, were down to once a week, but, even after all this time, she still had a job to keep him down to two or three times a week. Even then she reckoned he masturbated if she ever left him alone in the house.

That would be no good for her, her friend said. She wouldn't be able to cope with having sex that often. Fortunately, her partner also had a low sex drive – they were lucky if they managed it twice a month now. It had been once or twice a week when they first started living together, but it had soon declined. As for masturbation – they had never talked about it. She wasn't even sure he'd know how to. The taller woman laughed and then expressed her amazement that they had never talked about it. One of the first things her partner would tell her when she returned after leaving him alone in the house was whether he had masturbated. Maybe he didn't always tell her, but she thought he did usually.

They paused to sip their drinks. The other woman hesitated, summoning the courage to ask what for her was a big question. Falteringly, in a quiet, confidential voice, she asked her friend how often she came during sex. Without any surprise or self-consciousness, the taller woman replied that she came sometimes – but her partner expected her to come every time. In the early days, she had told him once or twice that she hadn't and he had sulked for hours. Ever since, if she didn't come, she faked it. Besides which, she said, he could keep going for hours, just waiting for her. If she didn't fake it, he would never stop.

Feeling uncomfortable, the other woman took another sip. She found some consolation in her friend's answer, but not much. There was a brief silence. She knew it was her turn to reciprocate but couldn't decide whether to tell the truth or not. In the end, she admitted that she had never come during sex. In fact, she didn't think she had ever had an orgasm. When they had sex, her partner just stuck it in and ejaculated. Admittedly, he had tried – at least in the early days. He used to fiddle around between her legs during foreplay, but it had never really done much for her. If anything, she used to get embarrassed rather than excited. In the end, she had told him not to bother and just to get on with it. Once or twice she had felt something during intercourse but it had never amounted to anything. She wasn't sure whether it was her or him, but she had just never got anything from sex – except their baby.

In a matter-of-fact way that shocked her friend, the taller woman asked if it was any better with other men. The other woman smiled and shook her head. There hadn't been any other men, she said. The taller woman asked if she meant never, or not since she had been living with her partner. When told never, she expressed disbelief, astounded that anybody could get to thirty and only ever have had sex with one man. *She* must have had at least *twenty*. When asked if she meant when she was younger or if she had had other men since living with her partner, she laughed at the innocence of the question and said both. She expected at least one fling a year. She'd even had sex with someone else while she was pregnant with her first child. The thought of only having sex with her partner and missing out on the excitement of someone new occasionally was unimaginable.

Unsure of how she felt about her friend in the light of these revelations, the other woman tried to think of a suitable response. In the end, she simply said she didn't know how the other managed to get away with it. Her own partner hardly ever let her out of his sight and she was sure he would know if she even thought about being unfaithful. Her friend said she sometimes wished *her* partner were a bit more attentive. Not too much, but a bit. Sometimes she thought he didn't care what she did. He was never around when she needed him – for all she knew, he could be having a different woman every week. There were always women around him. He could take his pick. It would be a hopeless task trying to turn him into an attentive partner and father. In any case, if she did, it would give *her* far less chance to enjoy herself.

She leaned forward and touched her friend's hand. Apparently taking no account of the latter's confessed fidelity, she told her what she should do if she wanted a partner who would give her orgasms. In a whisper, she confided that the best lover she had ever had had been gay. She hadn't realised until she had seen him holding hands with another man in a bar a few weeks after they had finished. He had been wonderful, and had always seemed to know exactly what she wanted. Try and find yourself one, she urged.

Her friend was just about to recoil at a prospect so alien to her, when one of the men at the bar began to make his way unsteadily over to their table. He put down his drink and balanced himself awkwardly with arms outstretched, fists on the table. Dribbling slightly, he observed that the two women looked as though they knew how to have a good time. Why didn't they come with him? He could give them a night to remember. Maybe they should toss a coin to see who could have him first.

The taller woman told him to drop dead. When he didn't leave, she stood up and pushed him away. He fell backwards on to the floor, then got to his feet and swore at them before staggering back to the bar. The woman sat back down, picked up her drink and smiled at her friend. Deep down, she observed, men were all the same. Drunk or sober, young or old, they were only interested in one thing. If they had half a brain as well as genitals they would really be quite dangerous.

~

Sexually, men *are* much more similar to each other than are women. Virtually all men ejaculate (whereas not all women have orgasms). Virtually all men masturbate (whereas nearly a quarter of women do not). Virtually all men have nocturnal orgasms at some time in their lives (whereas 60 per cent of women do not). Nevertheless, men do still differ in the ways they pursue reproductive success. There are roughly four different strategies.

One strategy mentioned in Scene 35 is bisexuality (discussed at some length in Scene 30). The two other strategies described by the women in the scene represent the two ends of the male spectrum of sexuality. We have met these two types of men before in a different context (Scene 19). One specialises in sperm warfare, the other in avoiding sperm warfare. In between these two types lie the majority of men who intermesh an avoidance and a seeking of sperm warfare in as productive a way as possible. Just which strategy a male is programmed to adopt will depend largely on his rate of sperm production – which in turn depends on the size of his testes.

Men have a pair of testes of unequal size (on average, the right is 5 per cent larger), which hang in a scrotal sac at different heights (more often the left is lower). The testes of all mammals originate inside the body in the same position as the ovaries – and in many species that is where they stay. In other species, such as humans, they descend into the scrotal sac before birth and stay there throughout life. In yet other species, the testes descend during the breeding season, then go back inside the body for safe keeping once the breeding season is over.

Scrotal testes are more vulnerable than internal testes and can easily be damaged. The main compensation is that, because they allow the sperm to be stored at a lower temperature than if they were stored inside the body, it is easier for the sperm to keep fitter and healthier for longer. When men are naked, their sperm are stored at a temperature that is 6°C cooler than if they were inside the body, but when they are clothed, the difference is only 3°C.

On average, taller and heavier (but not obese) men have larger testes. Some men, though, have testes that are relatively large for their body size; others have testes that are relatively small. This difference is genetic and heritable. As long as there are no associated clinical problems, even the smallest of testes can produce enough sperm for fertilisation in the absence of sperm warfare. Moreover, small testes are less vulnerable and less likely to be damaged than larger ones. So why don't all men have small testes? The answer is that when sperm warfare is likely, small testes are a major handicap. The sexual strategy that a man does best to pursue, therefore, is dictated to a large extent by the size of his testes.

Men with larger testes manufacture more sperm per day, ejaculate more often, and introduce more sperm at each intercourse. Interestingly, they don't ejaculate more sperm during masturbation. They spend less time with their partner, and are more likely to be unfaithful and to choose a partner who will also be unfaithful. The converse is true, in all these respects, for men with smaller testes.

In short, men with larger testes are programmed to specialise in the pursuit of sperm warfare – warfare which, because of their large sperm armies, they are likely to win. Men with smaller testes, on the other hand, are programmed to specialise in mate guarding, fidelity and the avoidance of sperm warfare – warfare which, because of their small armies, they would be likely to lose. So who is the more successful reproductively, a man with small testes or a man with large ones? The answer seems to be neither. Just as for bisexuality, evolution seems to have produced a balance such that men with large testes and men with small do equally well.

To illustrate this point, imagine a population of men with small testes, injecting few sperm into their respective partners and making no attempt to inseminate other men's partners. Into this population comes a man with large testes who not only claims a partner of his own but also tries to inseminate other men's partners. At first, he does extremely well. Every time he inseminates another man's partner he is likely to win the sperm war because he introduces a larger sperm army. Yet, at the same time, he is safe from the possibility of 'cuckoos' appearing in his own nest because other men are not inseminating *his* partner. As a result, at each generation men with large testes produce more children than men with small. Moreover, their male descendants inherit their large testes, their promiscuity and their ability to win sperm warfare.

But this success eventually becomes self-defeating. At each generation there are more and more men with large testes – descendants of the original invader – and in the end such men no longer have an advantage. First, they are no longer assured of winning sperm wars because the women they inseminate are being inseminated by other men with large testes. Secondly, their own partners are now vulnerable to being inseminated by other men with equally large testes. Thirdly, the greater promiscuity in the population puts everybody at greater risk to disease, particularly the most promiscuous – such as themselves. Thus, when there are too many men with large testes in the population, the less well-endowed ones who concentrate on guarding their partner against other men actually do better. Not least this is because such men are less at risk to disease and their smaller testes are less vulnerable to accident and damage.

So if sperm war specialists with large testes ever become too common in the population, they actually do worse reproductively than men with small testes. We have met this situation before (Scene 30) and the outcome should be the same. The proportion of men with large testes settles down at the level at which, *on average*, such men do no better and no worse than men with small ones.

In Scene 35, the man with large testes was raising two children who may or may not have been his. In addition, he may or may not have produced other children with other women. The man with small testes was raising one child, which was certainly his (according to his partner). The latter male has greater certainty, the former has greater potential. On average, however, the two types of men should produce equal numbers of children.

In between these two extremes of testis size and reproductive strategy lie the majority of men – those with testes of intermediate size. These employ a 'mixed' strategy by which they try to strike the best compromise between mate guarding and sperm warfare, but specialise in neither. The presence in the population of this 'mixed' majority might seem to complicate the picture, but in fact the conclusion remains the same – their proportion is also fixed at the level at which on average their reproductive success is no better and no worse than males with smaller or larger testes. In effect, as long as a man pursues a reproductive strategy appropriate to the size of his testes and his rate of sperm production, he should on average do just as well as other men with different-sized testes.

The sperm war specialist in Scene 35 had a large penis as well as large testes, whereas the mate-guarding specialist had a small penis as well as small testes. This should not be surprising, given that the penis does have a role in sperm warfare – the removal of any seminal pool from the vagina (Scene 21). On the whole, however, a large penis is not as consistent a feature of sperm war specialists as large testes – because, compared with testis size, penis size influences the outcome of sperm warfare far less often. After all, penis size is important only on those rare occasions when a man has sex with a woman very soon after another man – so soon that the latter's seminal pool is still at the top of her vagina. In contrast, testis size, via sperm number, is *always* important in sperm warfare.

This explains why there has been much more pressure on sperm war specialists to have larger testes than to have a larger penis, but it does not explain why any man should have a small penis. There is, of course, a lower limit to penis size – that below which a man cannot introduce sperm far enough up the vagina. There is also an upper limit – that beyond which a man would be unable to thrust without damaging the woman. Within this range, though, why should any man have a penis smaller than the upper limit? The answer is that within this range there is no real disadvantage to a smaller penis (other than being an infrequent handicap for sperm warfare) and there may even be an occasional advantage.

On the one hand, a smaller penis is no disadvantage in terms of sperm retention. First, it is not particularly less efficient at delivering a seminal pool to the top of the vagina. This is because even when a penis of below-maximum size withdraws after insemination, the vaginal walls close behind it (Scene 3), effectively pushing the seminal pool to the top of the vagina. Secondly, penis size has no influence on the probability of a man's partner having an orgasm during intercourse.

On the other hand, a smaller-than-maximum penis may even be an advantage, particularly during routine sex. Everything depends on the costs and benefits of a man removing *his own* seminal pool whenever he inseminates a woman twice in rapid succession, say within thirty minutes or so (Scene 25). On these occasions, a smaller penis will be an advantage if the previous seminal pool is better left in place. If the pool is better removed, a smaller penis can still do the job – it just takes longer. In many ways, therefore, a smaller-than-maximum penis endows a man with greater flexibility than does a larger one. It is also less likely to be accidentally damaged.

Testis size and penis size vary not only between males *within* a population, they also vary *between* populations and races. On average, even relative to body size, Negroes tend to have larger genitals than Caucasoids, and both have larger genitals than Mongoloids. The number of sperm inseminated during intercourse varies accordingly. It has been claimed that these differences between populations reflect different balances in sexual strategy, just as they do between men within populations. In other words, populations with larger genitals (on average) should contain more men pursuing sperm warfare than populations with smaller genitals (on average). Although this claim has not been tested for different populations within the human species, it has been tested *between species*.

In some species of primate, such as chimpanzees, females often mate with several males and nearly every conception involves sperm warfare. In others, such as gibbons, females rarely mate with a male other than their partner and conception rarely involves sperm warfare. Linked with this difference, chimps have much larger testes relative to their body size than do gibbons. Humans, with 4 per cent or more of children conceived via sperm warfare (Scene 6), are intermediate between the two, both in terms of warfare risk and testis size.

Not only primates but many other groups of animals have been shown to have testes of a size appropriate to the risk of sperm warfare. From butterflies to birds and from mice to men, the more likely a male's sperm are to engage in conflict, the larger his testes relative to the size of his body.

SCENE 36

Exquisite Confusion

His moment had come. It was a year since his second partner had left him, a year of celibacy, anticipation and nothing but masturbation. But, unless he did something totally stupid, it looked as though this was going to be it.

It wasn't the first party he had thrown since she had left. It was an easy way of filling the house with lots of young women (and a few men). But it was the first at which he had managed to keep anybody's attention for more than about ten minutes. He had targeted this young girl, nearly thirty years his junior, the minute she walked in the room. A friend of a friend, she stole his attention immediately. He had waited until she looked drunk, then moved in. They had been talking for hours. Even when she had gone to the toilet, always a dangerous moment, she had come back to him. For the last hour, their conversation had been about sex. With an honesty born out of drunkenness, they had traded increasingly personal details, such as how often they each masturbated and how many partners they had each had. She was well into double figures and sometimes despaired of herself, she confided, but she could never really think of a good reason to say no.

She began to look sad and emotional. He said she shouldn't think badly of herself. Without her experiences, she wouldn't be the person she was now: so worldly, so relaxed and so attractive. Those past sexual experiences had surely helped her become the person that he, for one, was now finding so irresistible. With that, he had stroked the side of her face with the back of his fingers. She leaned her face against his hand and small tears appeared in her eyes. She smiled and apologised as he wiped the tears from her cheeks. Then, as he continued gently to touch her face, tracing his fingers round her eyebrows, nose and mouth, she looked him in the eyes with an expression he recognised from a previous lover – she wanted sex. There were still people in the room, but the pair of them were now oblivious. He moved his hand from her face to her neck, then back again, stroking with his finger tips. Then he moved down and traced round the edge of her low-cut dress, briefly slipping his fingers inside so that they brushed over an erect nipple. She made a tiny noise in her throat. His moment had come. He took hold of her hand and suggested they go somewhere more private. Unhesitatingly, she let him lead her.

As he led her upstairs to his bedroom he congratulated himself. Maybe, at last, he was becoming competent at the sexual arts. The thrust of the conversation he had just used to interest this girl had been almost identical to the one he had used when he first seduced his last partner, ten years ago. In fact, apart from this girl's relatively tiny breasts, she could almost *be* his last partner.

Within seconds of bolting the door, they were lying naked on the bed and he had started the routine that according to his ex-partner had made him an exquisite lover. His hands, lips and tongue began to move over her body, with varying degrees of lightness and strength. It was a well rehearsed routine that had never failed to stimulate her to orgasm when she was as aroused as this girl seemed to be. She used to lie there as he caressed her, getting more and more excited as he spent more and more time massaging her clitoris until she eventually climaxed. However, as he worked slowly over the young girl's body, judging the right moment to move to her clitoris, he realised that she wasn't reacting. Then suddenly she seized him by the neck and kissed him passionately, almost violently. Taken aback, he did his best to respond before trying to take the initiative again. Rolling her on to her back, he forced his hand between her legs. For him it was an unfamiliar angle, and she was moving so much he couldn't find her clitoris. He had rarely had such a problem with his ex-partner. But then she had always lain still while he stimulated her.

Thrown from his usual routine, he placed his fingers in the girl's vagina instead. She was very wet. But instead of allowing him to gently move his fingers up and down, in and out of her vagina, as did his ex-partner, she responded by vigorously rubbing her pelvis against his hand for a few seconds. Then she quickly manoeuvred him on to his back and herself up his body. Next, in a movement he had never come across before, she seemed to grab his penis with her vaginal lips and virtually suck him in. He felt awkward lying on his back because he was used to being on top. He tried to start thrusting but the girl was thrusting too. His ex-partner used to lie still and leave him to orchestrate the whole intercourse. But then, she had usually climaxed during foreplay, if she was going to climax at all.

Concentrate though he might, he just could not synchronise his thrusting to the girl's. He kept coming out. Each time she lost his penis, she would simply relocate it with her vaginal lips and suck him in again. In the end, he gave up trying to thrust and left all the movements to her, patiently and politely waiting for her to climax before he ejaculated. The sounds in her throat grew louder and louder. Then, suddenly, she stopped – no contractions, no apparent release or relief, no anything. As she stopped moving he began to thrust, but he was too late. Without waiting for him to ejaculate, she removed herself and lay breathlessly on the bed, leaving him erect, wet and frustrated.

As she lay, recovering, he kissed and stroked her. Desperate to inseminate her, he tried to keep her interest, but it didn't work. A few minutes later, she got off the bed and started to dress. He assumed she was going to the toilet, but as she got to the door she said goodbye. She had really enjoyed it, she said, but she had to go. He protested that he hadn't yet come. She told him to do it himself – if she had more time she would help him, but she really had to go. Her boyfriend was coming to pick her up and take her home. As she opened the door, he suggested they should do it again sometime. She smiled, shook her head, and went.

After dressing, he went back down to the party, but most people had gone and none of the girls still there attracted him. He ushered the last few people out and went back to bed. But he couldn't sleep, even after masturbating. He couldn't believe he had missed his chance. He was nearly fifty and still he couldn't guarantee making the most of his opportunities. It wasn't that he was inexperienced – just that he had never met anybody like this girl before.

He had had little opportunity to learn about sex during adolescence, and hadn't really got to grips with intercourse until he had met his first partner. During their fifteen years together they had produced two children. Yet not once in their entire relationship did she have an orgasm. She had been totally passive. She didn't like him touching her between her legs and refused to allow his face anywhere near her pubic hair. Because of their lack of foreplay, she was rarely wet when he tried to enter her. To get herself lubricated, she used to urge him to put the tip of his penis just inside her vagina and then work himself backwards and forwards gently. Once he was fully inside, she then liked him to ejaculate quickly. If intercourse lasted too long, it was as if she became embarrassed. On more than one occasion, she had actually made him stop and withdraw because he was taking too long. Ironic though it seemed in hindsight, he had convinced himself while he was with her that there was no such thing as the female orgasm.

When he was thirty, he had his first opportunity to be unfaithful. At a party at work, one of the junior girls had started stroking his leg with her foot. They went back to his office and lay uncomfortably on the uncarpeted floor. Assuming she would want the same as his partner, he missed out foreplay completely and, in a tangle of half-removed skirt and knickers, trousers and pants, tried to penetrate her immediately. She was offended by his lack of consideration and disappointed that she wasn't going to get an orgasm. Suddenly she had seemed to become aware of the discomfort of the situation and equally suddenly became overtaken by guilt – she, too, had a partner at home. Pushing him away, she said that maybe sex wasn't such a good idea after all, and left.

He nearly suffered the same fate three years later when he tried to make love to his first real mistress. She, too, took offence at his initial urgency. But instead of dismissing him after the first occasion and despite being ten years his junior, she set about educating him – and there was a lot he needed to know. For example, in all his time with his partner he had neither touched nor seen her clitoris. In truth, he didn't even know she had one and, with the benefit of hindsight, he suspected that neither did she. Two things his mistress did share with his partner were a mild liking for having her nipples sucked and a strong dislike of having his face between her legs. Whenever he tried, she restrained him. He never did in fact *see* her clitoris. However, she managed to show him both how to find it with his fingers and, having found it, what to do with it. She didn't like him to touch the clitoris bulb directly, preferring him to massage and squeeze the area around it. But she never wanted to climax during foreplay. He became expert at judging the right moment to move from using his fingers to using his penis. And as long as he judged the moment correctly, she always seemed to climax during intercourse. Somehow, though, he always had the feeling that she was faking it.

This first infidelity lasted a year and passed undetected by his partner. It ended when his mistress met someone nearer her own age. A year later, he began his second affair. He was now relatively experienced at the deceptive techniques of infidelity, but was still not fully convinced that the female orgasm was a reality. As it turned out, his second mistress was not the person to enlighten him.

Most of the time he was with her he felt like her puppet. They worked together and from time to time had legitimate reason to spend time together away from their respective partners. Eventually, she let him into her bed on condition that he did not try to penetrate her. On their first night, she helped him to ejaculate in exchange for having him stroke her body and, particularly, to place his head between her legs and lick her. At first he had no idea what he was doing. Though he was now thirty-five, she was the first woman who had actually urged or even allowed him such intimate contact. Yet she was virtually insatiable, wanting him to lick her genitals, thighs and anus for twenty minutes at a time so that she was covered in his saliva. Even so, she never had an orgasm or even became excited; she seemed to find his attention therapeutic rather than stimulating.

A year later, on what was their tenth assignation, she was still insisting that he could only have his excitement as long as he did not penetrate her. This time, their business was abroad and they were away for a fortnight. For two nights, he let her tell him what to do but on the third night, as they lay naked and he had his head between her legs, his excitement and frustration at last took him beyond the threshold of reason. He slid up her body, kissing her as he went. Then, while kissing her face and mouth, he slipped his penis into her vagina. She immediately struggled but he pinned her down, smothering her protestations with his mouth and free hand. She became violent, hitting and scratching him whenever she could free a hand. But he was too strong for her and so excited that he ejaculated quickly. When it was over, she called him a bastard – but didn't throw him out. Nor did she get dressed, and within an hour she was taunting him into forcing her to have sex again. Over the remaining week, they had sex every night. Each night, he had to find a different way of restraining and forcing her. She conceived that week but three months later miscarried when her partner, having been denied sex for well over a year, decided to leave her rather than bring up somebody else's child.

Until his mistress miscarried, our man had been tempted to leave his partner and help her raise the child. After the miscarriage, though, he could not face the prospect of years of having to force sex, and finished the relationship. But it was too late. His partner had been told of his affair by his mistress, and three months after finding out she left him for another man, taking their two children with her.

There followed three short relationships, each girl as different from the others as she could be. The first was afraid of sex. Raped as a young girl, she had a fear of the erect penis that distressed him almost as much as it did her. They drank excessively. Over the course of a year they shared a bed perhaps five or six times, but never managed to have intercourse. He didn't realise her similarity to his last mistress until, at their last meeting, she asked him to rape her. He obliged, emotionally torn at watching her distress as he forced his penis inside her. Unlike his last mistress, however, she did not come back for more and he never saw her again.

Now nearly forty, his next experience was a great shock. He had been out celebrating with a small group of friends from work. They had all returned to the house of a young couple barely in their twenties, and were soon plunged into a drunken party with very loud music and much raucous laughter. Almost immediately the party was under way, the female host had disappeared to her bedroom. She returned a few minutes later wearing a short, almost transparent, white dress. Then, seemingly unnoticed by her very drunk partner, she openly took the hand of one unattached man after another and led him away to her bedroom. When it came to his turn, she told him she had been saving him till last. She took him upstairs and, without locking the door, removed his trousers and underwear. As she took off her knickers, he could smell the semen from the previous men.

Sitting on the edge of the bed, she began to give him oral sex. From the noises she was making, she was becoming excited even faster than he was. Eventually sounding frantic, she pulled him on top of her and scrambled with her hands to insert his penis. As he pushed himself in, she let out a blood-curdling scream and her body began jerking wildly. The violence of her reaction was such a shock to him that his erection began to fade. He carried on thrusting, trying to rejuvenate himself. But just as she was building up to a second crescendo and his own ejaculation was only seconds away, her partner burst into the room. Only an instant before he ejaculated he was pulled off her, pushed to the ground, and ordered out. With apparently little surprise and with minimal aggression, her partner proclaimed that now it was his turn. Hardly faltering in her build-up at the change of penis, the girl released yet another blood-curdling scream.

His third experience was with an exotic woman with whom he had sex about five times. The first two times were clearly a disappointment to her. She had refused to let him enter her without extensive foreplay. Yet the foreplay he provided, borrowed from his technique with his first mistress, seemed to do nothing for her, and on each occasion it was with reluctance that she eventually let him have intercourse. The third time, she told him she would do it herself. She masturbated in front of him and, after she had climaxed, allowed him to have intercourse. Years later he was to realise that, had he already experienced the woman who was to be his second long-term partner, he would probably have been able to excite this woman. What she did to herself was precisely what that second partner later taught him to do to her. After twice watching his exotic lover masturbate, he began to feel fairly incompetent. In the end he found himself wondering if what she really wanted was for him to force her, like his second mistress. He tried this, failed, and was thrown out, never to be invited back.

Soon afterwards he met his second long-term partner, twenty years his junior (the woman who had just left him, taking their child with her). The first time they had sex she was so upset by his urgency that he almost forfeited his second chance. Instead, she relented and taught him what she needed. After two or three years of education he became quite proficient. Quiet and sensuous, she liked him to stroke her gently while she lay back, soaking up the sensations. Unlike his first mistress, however, sucking her nipples did little for her. She showed him precisely where her clitoris was and how to find it. She liked the bulb itself, once lubricated, to be rubbed and massaged directly.

Eventually, he could virtually guarantee giving her an orgasm during foreplay, as long as she wanted him to. Rarely, very rarely, did she have or even want an orgasm during intercourse, saying they were so weak they weren't worth having. When each episode began, she would tell him whether she felt like an orgasm during foreplay or whether she just felt like intercourse. If she changed her mind or if for some reason he was unable to excite her, she would openly masturbate to satisfy herself. And she told him about her nocturnal orgasms. As far as he knew, none of his other partners had experienced them.

After ten years with her, he had felt able to excite and satisfy any woman. Tonight's events had proved him wrong. Shocked and disappointed, he drifted off to sleep in a sea of confused thoughts. He would never understand women.

~

Compared to men, women are incredibly varied in their sexual characteristics. Some women (2–4 per cent) never orgasm; others

(5 per cent) have multiple orgasms, scarcely coming down from one climax before building up to the next. Ten per cent never climax during intercourse; another 10 per cent nearly always climax. Some prefer to be totally passive in their build-up to climax during intercourse; others prefer to be more active. Fifty per cent masturbate regularly; 20 per cent never do. Finally, 40 per cent have nocturnal orgasms, whereas others cannot even imagine such an event.

Added to, and often associated with, these differences in orgasm pattern are differences between women in the sensitivity of different parts of their bodies. Some women have sensitive nipples; others do not. Some have a clitoris so sensitive that they dislike the bulb being touched; others have one that responds very little to stimulation.

It is not quite true that no two women are the same – but to many men it often seems that way. Why *is* there so much variety to female sexuality, and what bearing does such variety have on the female pursuit of reproductive success? The answer is that it is yet another manifestation of the advantages gained from confusing men (Scene 2) and testing their abilities (Scene 27). It works because, as we have seen before (Scene 27), men have to learn their sexual technique.

The man in Scene 36 was bemused. However, he wasn't quite the failure he imagined himself to be. In fact, reproductively, his life was fairly successful. He had two children with his first partner and one with his second (assuming they were all his). But he could have had even more. At different times during his life he had the opportunity to impregnate seven other women, and to varying degrees he had failed each time. On some occasions he failed even to inseminate. On others, he failed to handle the relationship either well enough or for long enough to have a realistic chance of paternity. The nearest he got to a child with a further woman was with his second mistress. Eventually, her body allowed him to fertilise one of her eggs. But it took him a year of missed opportunities, and that was just too long. Had he impregnated her earlier, he might successfully have deceived her partner into raising another's child. Then the baby might never have been miscarried and his reproductive success might have been 33 per cent higher than eventually it was.

His failure to capitalise on opportunities to reproduce with these seven women was due in every case to some combination of surprise, shock and misjudgement. He did what he could, but he was confused by the sheer variety of female sexuality. As he moved from woman to woman he tried to extrapolate from his past experiences. He assumed that techniques of seduction, stimulation and insemination that had been successful in the past would be successful in the future. To some extent, his approach worked – his gradual accumulation of experience with different females slowly increased his awareness and competence in the sexual arts. But, even after encounters with eight very different women, he still failed to inseminate a ninth.

In some species, experience gained by a male with one female can readily be extrapolated to another. In other species, however, such extrapolation is less successful – and it is least successful in those such as humans in which the underlying female strategy is the confusion of males. Then, not only do individual females confuse their partner through unpredictable swings of mood and behaviour (Scene 2), but also the female population as a whole manages to confuse through variety. As was clear from Scene 36, a woman gains in three main ways from being as different as she can be from other women.

First, she can set each man a much more challenging test of experience and competence (Scene 27). From this she can quickly learn whether he has a little or a lot of experience of other women. Because women are so different, a man will only have encountered her type before and thus know how to treat her if he has had a lot of experience. If her priority is to avoid disease or to find a faithful partner, she may actually prefer a naïve man. On the other hand, if her priority is to find a man who is genetically more attractive (Scene 28) she may prefer someone whom many other women have found attractive (Scene 27).

Secondly, she has more initial control at each sexual encounter with a new man, which lasts for as long as it takes the man to fathom the best way to relate to her. The result is that she has more time to assess the man as a long-term partner. In Scene 36, three women tested the main character sexually, but managed to avoid being inseminated before rejecting him. One took longer to decide how she felt about him, but was able to keep both the number of inseminations and sperm retention down to a minimum in the process. Two others managed to avoid insemination for nearly a year while they made up their minds. One finally rejected him, and the other conceived but later miscarried.

Thirdly, after accepting a man as a partner, she can still educate him in her needs without facilitating his infidelity. In the scene, two of the women (his first mistress and his second long-term partner) evidently found the man's other qualities attractive enough to tolerate his initial sexual incompetence. They then began the relatively long process of educating him in helping them to climax (Scene 27). This could have been self-defeating, if his increased competence had made it easy for him to seduce other women. Of course, it did help him a little, but not as much as it would have done if women were more similar. His first partner pursued the ultimate version of such a strategy. By not having orgasms herself, she could not educate him at all. As a result, she made him miss one opportunity for infidelity and slowed him down in taking advantage of another.

Although it is not difficult to see that women can benefit considerably from being different from one another, there has to be a limit to this range of possibilities. Different though women may be in their individual sexuality, therefore, we can nevertheless still recognise a number of broad categories. These categories are probably genetic, are probably in some sort of balance in the population, and are probably all most successful if they are not too common.

We have met evolved balances of genetic categories before in earlier scenes. The proportion of bisexuals and heterosexuals (Scenes 30 and 31) and the proportion of men with different-sized testes (Scene 35) are both examples of different categories co-existing in an evolutionary balance. When too common, each category has a success rate that is below average. When rare, each has an above-average success rate. The incredible variation in female sexuality is simply the most complex example we have yet encountered of such a balance. As in all previous examples, our expectation will be that the balance point has been fixed by evolution such that reproductively all of the different categories do equally well.

So, what are these different categories? More importantly, are the categories different simply for the sake of being different, or is each category linked to other aspects of sexual behaviour? The latter seems the more likely. Just as large testes (Scene 35) and bisexuality (Scene 30) predispose some men to specialise in particular sexual strategies, the different orgasm patterns shown by different women will also predispose them to particular sexual strategies. For example, some orgasm patterns will help the women concerned get the most out of the promotion and exploitation of sperm warfare. Others will help the women concerned to get the most out of fidelity and an attentive partner. Very crudely, while accepting that the tremendous variation in female sexuality largely defies categorisation, we can make life easier for ourselves by recognising four main types.

The first consists of women who are programmed sometimes to have (and sometimes not to have) the full range of possible orgasms (masturbatory, nocturnal, foreplay, intercourse and post-play – with some being multiple). Although only about 5 per cent of women have absolutely the whole range, an additional

25 per cent have all except multiple orgasms, and an additional 40 per cent again have all except multiple orgasms and nocturnals. All women in this category can manipulate sperm retention in all of the ways we have discussed (Scenes 22 to 26). In particular, they can vary the frequency with which they have and do not have every type of orgasm so as to take maximum advantage of the men and opportunities which present themselves. They are thus probably the women most able to exploit sperm warfare whenever it is advantageous to do.

Although as a whole this first category of women is probably the largest (about 75 per cent), there is tremendous scope for women to differ *within* the category. For example, about 30 per cent within the category both masturbate *and* have nocturnals, whereas 50 per cent *only* masturbate and 10 per cent *only* have nocturnals. These differences probably reflect variety for variety's sake, as we have discussed. Masturbation and nocturnals are alternative means to the same end (Scenes 22 and 23). Variation in emphasis on one or the other outlet allows women to differ. It may also influence their scope for secrecy in preparation for infidelity (Scene 23), but it is unlikely to have major repercussions on their ability to manipulate sperm retention. In addition to differences *between* women within the category, each woman also has so much scope for varying her own behaviour (from phase to phase and man to man) that she need never become predictable.

The second category consists of women who are programmed to avoid one or other of the major avenues for the manipulation of sperm retention and warfare. Either they do not have solitary (masturbatory and nocturnal) orgasms or they do not climax in the presence of men. In most cases, however, such women sacrifice only a little of their manipulative ability. In compensation, they gain by being slightly less common. Thus, about 10 per cent of women are programmed neither to masturbate nor to have nocturnals, which has the effect of delegating responsibility for the manipulation of sperm to orgasms in a man's presence. The full range of sperm manipulation is still possible (via orgasms during foreplay, intercourse and post-play – Scenes 22 to 25), but it becomes much more important to the woman that she chooses a sexually competent man. Other women (again about 10 per cent) are programmed never to climax in a man's presence but do masturbate and/or have nocturnals. They also have the full range of sperm manipulation. They can secretly prepare a strong or a weak cervical filter – but what they cannot do is change their minds at the last minute and bypass an over-strong filter (Scene 25).

The third category consists of those 10 per cent of women who climax virtually *every time* they have penetrative sex. Such women can still manipulate sperm retention, and thus play an active part in influencing the outcome of sperm warfare – as long as they vary the timing of their climax relative to the moment of insemination (Scene 25). An orgasm more than one to two minutes before the man allows low sperm retention; any time thereafter allows high sperm retention.

The final category consists of those 2–4 per cent of women who never orgasm, either during intercourse *or* via masturbation *or* nocturnally. In compensation, as mentioned above, such a woman gains by not educating her partner in sexual techniques that would aid his infidelity. Unlike those in the previous three categories, such women are the least able to influence the outcome of sperm warfare. They are, however, well suited to the relatively safe process of selecting a mate and making the most of that relationship. They do have some control over sperm numbers, largely through waiting for their cervical filter to weaken with time since their last intercourse (Scene 22). But since sperm numbers are far less important for conception than they are for warfare, in the absence of any possibility of the latter it does not really matter how many sperm such a woman retains from each intercourse.

Although it is possible to retain *too many* sperm for conception from a healthy ejaculate (Scene 19), it is much more difficult to retain *too few*. Admittedly, ejaculates containing few sperm are often sub-fertile, even if the woman retains them in full. But this is not because they contain few sperm – even vasectomised men who occasionally introduce tiny numbers of sperm (perhaps fewer than a hundred) into women have fathered children. It is rather because the clinical conditions that cause some men to introduce few sperm (perhaps only tens of millions instead of hundreds of millions) at the same time render their ejaculates sub-fertile.

The women in this final category do, of course, lose the benefits of orgasm as a defence against infection. But then, of all the categories of women, they are also the least likely to encounter sexually transmitted diseases. In any case, it seems that even women programmed to avoid manipulating sperm via orgasm may nevertheless begin to orgasm if they are ever exposed to genital infection. Just how they do so will depend on their anatomy. If their clitorises are sensitive to stimulation they may suddenly experience the urge to masturbate. However, it is often part of the programming of such women that their clitorises are relatively insensitive to stimulation. This may make it difficult for them to masturbate to orgasm but, of course, in no way prevents them from climaxing nocturnally. Clitorectomised women, for example, may still have nocturnal orgasms.

The predisposition of different women for different sexual strategies adds a final nuance to mate selection – an element that can be added to those we have already accumulated (Scenes 18, 20, 21, 27, 28). This is that a woman should seek a partner whose approach to sperm warfare complements her own. The small-testicled man in Scene 35 who specialised in mate guarding was a good choice for the woman who never climaxed. Similarly, the large-testicled man in the same scene was a good choice for the woman with a varied orgasm pattern who could gain the most from infidelity and sperm warfare. This woman gained in two ways. First, her partner's specialisations for sperm warfare meant that his large sperm armies gave the armies from her other lovers a good run for their money. Secondly, because he pursued infidelity himself, he often left her free to pursue her own reproductive success by following the same course.

The final point to stress is the one we have stressed in other scenes in this chapter – *on average* the different categories of women should all reproduce equally well. The proportions of each category in the population will have stabilised at precisely the appropriate level for this to occur. Consequently, no woman should be hindered in her pursuit of reproductive success by the orgasm regime with which she has been programmed – as long as her choice of mate and subsequent behaviour are compatible.

11

Final Score

SCENE 37

Total Success

The old man reached out a gnarled and wrinkled hand. His stiff fingers touched the hand of the woman sitting next to him outside their home. She turned her head towards him, looking blindly through opal eyes, and smiled. Eventually, everybody had come, some of them travelling for days to be there. For the first time in decades, the old couple were surrounded by their whole family, the product of their long lives together. Including themselves, there were five generations present. Their eldest son was now nearly seventy; their youngest great-great-grandchild, just two weeks old.

The old man looked at his partner's face, his eyesight as keen as hers was faded. The reality in front of him was a deeply wrinkled face, blind eyes that so cruelly imprisoned her, kind mouth and a one-toothed smile – she was so proud of that last tooth. But, for a moment, what he saw was the smooth, beautiful face, clear, dark, mischievous eyes, sensuous mouth and gleaming teeth that had so fired his senses over seventy years ago. He still remembered, as if it were yesterday, how he had chased her through the forest and, on a bed of sweet-smelling leaves, had explored and entered her taut and perfect body for the very first time. She had been a wonderful partner, mother and grandmother: the perfect companion throughout life. He could remember only one phase, as she approached menopause, when their relationship had been strained – but it had lasted a mere five years.

He had had many opportunities to be unfaithful, and at times had been tempted, but he had always resisted, so afraid had he been of disease and of losing his princess. His last chance had been in his early sixties when a flirtatious young girl of twenty, so impressed by his status and well-preserved body, offered him hers for the night. But he had turned her down. It was one of the best decisions of his long life, because five years later she was dead. She had succumbed to the sexually transmitted disease which had become the scourge of their community and which she had probably been harbouring on the night she offered herself. What would his partner and his family have done for the last twenty-five years had he died and not been around to sort out their problems and give them the benefit of his experience?

The night after his last chance for infidelity, he and his partner had sex for the very last time. Somehow, ever since, it had never seemed worth taxing their ageing bodies with the effort – they couldn't have been closer, sex or no sex. Now, swamped by the emotion of the occasion, he told his partner that she was looking wonderful.

Once again, the old woman smiled her one-tooth smile. Deep in her stomach, the pain that was now her constant companion was temporarily eclipsed by the sheer pleasure of hearing her family around her. She squeezed her partner's hand. It was ten years since she had been able to see him. No longer need she be troubled by the reality of his frailty and his recent deterioration. More and more, the image in her mind was not the worn old man she last saw with her failing eyes but the muscular young athlete she had known all those years ago. She still remembered, more clearly and accurately than him even, the first time she had surrendered her body to him as they swam naked in a lake.

It had been a few years before she conceived their first child, but after that it had been easy. She had never really felt tempted to be unfaithful. As her partner had grown in wealth and status, her biggest fear had been that she would lose him. There had been long spells in their life when she had found him irritating and arrogant, but he was always kind, considerate and fun. As their family had grown, not once had she found herself wishing for a better partner or provider. She asked him once again to tell her who was there – for a start, could he see all their children?

He looked around the huge group of people gathered outside their home. Not only their family but the whole village had assembled for the celebration, which was planned to last all day. The children were running and chasing, many laughing, a few arguing or crying. The adults were in scattered groups, some sitting on the ground, others standing. Most were drinking, eating morsels of food, and anticipating the main meal. Everywhere was the sound of conversation, laughter and occasional raised voices as people renewed old family friendships and feuds.

One by one, the old man located their five children. As he named each one he saw, she nodded. They were all there. She had had eight children but two had died in early childhood and a third, after having a family of his own, had died a few years ago in his fifties. Now the grandchildren, she urged him. He knew there should be twenty-three but couldn't remember all their names. He picked out the dozen he knew best, all in their forties, then had to rely on his partner to supply the names of the younger grandchildren, some of them still teenagers or younger. She remembered their names clearly, but that didn't always help as he couldn't even begin to remember what some of the younger ones looked like. Even so, when she told him a name, he said they were present.

When the list was complete, she sat back in satisfaction. Twenty-three surviving grandchildren. That was three more than her lifetime friend and rival who, after a complicated and colourful life of great sexual activity, had died ten years ago. Now, she missed their camaraderie. They had been lucky. Most of their generation had either died young from disease or accident or had been infertile, or had lost many of their children before they reached adulthood. Between them, she and her friend had populated a large proportion of the village.

The old man said he hoped that she wasn't going to ask him to count their great-grandchildren. She shook her head and laughed. He could see the oldest, now nearly thirty, with her new-born, their most recent great-great-grandchild. Last night, the old woman said, she and their daughter had worked out that there ought to be fifty-two of them and four more on the way, and already there were sixteen great-great-grandchildren. He sat back and relaxed, freed from his task of child-counting.

'Just look what we have done,' he remarked with satisfaction, not for the first time forgetting her blindness. She squeezed his hand again, then let him go, concentrating on the noise and conversation around her.

Suddenly, there was a commotion. A gang of naked children ran across the clearing as a group of young men, naked apart from a belt around their waists, emerged from the forest. On their shoulders the leaders were carrying the first of three large animals, already skewered on a wooden spit. The food for their feast had arrived.

~

In the course of this book, scene after scene has documented incidents of infidelity; moments and situations in which successive characters have snatched the opportunity to try to increase their reproductive success just that little bit more than their contemporaries. This final scene is a reminder that with the right person in the right situation, the best way to achieve the greatest reproductive success can sometimes be within a monogamous relationship.

The scene is also a reminder that long-term relationships, including monogamous ones, are estimated to have been a feature of human sexuality for around three million years. Our hunter-gatherer ancestors (Scene 16), from the savanna of Africa to the forests of South America and South-East Asia, from the aboriginal outback of Australia to the Eskimo heartland of Canada, almost always formed long-term male–female partnerships, most of which were monogamous. Unlike the relationship in Scene 37, these partnerships did not necessarily last a lifetime. Often a person had two, or even three, successive involvements. But each one was monogamous (give or take an occasional infidelity) while it lasted, involved deep personal ties, and lasted for several years – very similar to what happens in modern industrial societies.

Only during the fifteen thousand or so years of dependence on agriculture did many human cultures switch to polygamy. Women clustered around the men of greatest wealth – those with the largest areas of cultivated land and those with most livestock. Even polygamous relationships, however, are long-term and involve deep ties between the male and each of his females. The women are expected to be faithful to the man and he is expected to be faithful to his designated collection of women.

Not until the advent of urbanisation and industrialisation over the last few hundred years has there been a wholesale swing back towards monogamy, or serial monogamy. But even now women still cluster around the men of greatest wealth and status (Scene 18).

The woman in Scene 37 made a good decision, not to be unfaithful. If, as it seems, her partner was the best provider around and had the best genes available, she had nothing to gain reproductively from infidelity (Scene 18), and everything to lose (Scenes 9 and 11). She had only done a little better, however, than her close friend who had had a much more varied sex life. Presumably, this friend did not start with the advantage of such a good partner. Had she opted to be faithful, she might have been less successful – again, reproductively – than she was. How successful the two women were may have had more to do with their ability to attract a good mate than with their strategy of fidelity or infidelity. They each pursued the best strategy for their circumstances.

The man in the last scene also made a good decision in avoiding infidelity. He did have much to gain (Scene 13), but he also had much to lose (Scenes 9 and 11). In particular, in his community there was always a real risk of contracting a life-threatening disease, as well as the cost of losing his 'princess'. She was fertile, faithful and a good mother and grandmother. Any gains from infidelity might not have balanced the costs.

Although the majority of people do not pursue a lifetime of pure monogamy, most (with few exceptions – Scenes 19 and 30) achieve the greater part of their reproductive success within long-term relationships. Then, as long as the individual judges the moment and the situation accurately, infidelity, rape, group sex, prostitution, partner swapping and so on are all strategies that provide the opportunity to be just that little bit more successful, reproductively, than would be possible solely within a single relationship. But all carry risks if badly executed or if the individual does not have the weaponry of physique or character to carry them through.

Maybe the couple in the last scene, particularly the man, could each have achieved greater reproductive success through infidelity and sperm warfare – but in all probability, because of their circumstances, they would not have done. Their reproductive strategy of faithful monogamy had, *for them*, been a total success.

Even though the couple in Scene 37 were monogamous and faithful, they still could not escape the shadow of sperm warfare. It is irrelevant that neither of them *actually* engaged in sperm warfare. Their bodies will still have spent a lifetime 'on alert' for a war that never came.

Everyone's body – with no exception – is similarly alert. Continuously throughout reproductive life, the body both assesses the likelihood of sperm warfare and makes the appropriate preparations. When the likelihood is low, some preparations are made but they are minimal (Scenes 12 and 22). From time to time, however, absolutely every body contemplates behaviour that could lead to sperm warfare – and when it does, such contemplation triggers an escalation in preparation (Scenes 13 and 26). More often than not, the contemplation is mere fantasy and the preparations unnecessary – no war ensues. But even in our own generation, the majority of men at least once in their lives act on their fantasies and enter their sperm into battle – and, at least once in their lives, the majority of women do likewise and promote sperm warfare. When war does break out, each person's body will have done its best to make sure that it is well prepared and that there is a chance the war will work to its reproductive advantage.

Few people will have thought deeply about sperm warfare and its consequences before reading this book. Yet, if the phenomenon did not exist, human sexuality would be far less colourful. Without sperm warfare during human evolution, men would have tiny genitals and produce few sperm. Women would not have orgasms, there would be no thrusting during intercourse, no sex dreams or fantasies, no masturbation, and we should each feel like intercourse only a dozen times or so in our entire lives – those few occasions when conception was possible and desirable. Sex and society, art and literature – in fact, the whole of human culture – would be very different.

Over the millennia, sperm warfare has helped to shape society. In contrast, the last few years have seen two developments – one social and one scientific – which in their turn could change the face of sperm warfare. The *social* development was the emergence of child support agencies as governments tried to shift financial responsibility for the children of women without partners away from society and on to the absent father. Interestingly, in their attempts to avoid or defer their support, men everywhere have responded by invoking sperm warfare – expressing doubts about their paternity of this or that child. In the past, most such claims of non-paternity would have remained unresolved. However, the *scientific* development – genetic fingerprinting – has provided a relatively conclusive method of testing such claims (Scenes 16 and 30).

It is interesting to speculate how a world of child support agencies and paternity tests might change sexual strategy and the role of sperm warfare.

The main repercussions are twofold. First, men will find it more difficult to have sex and run (Scenes 13 and 29). No longer will it be as easy to deny paternity and leave child-raising to the woman. Secondly, women will find it much more difficult to trick men into raising children who are not theirs (Scenes 6, 8, 13, 16, 17, 18, 26, 31, 33 and 35). It is not difficult to envisage a future in which men will routinely pay for (or even be entitled to!) a paternity test at the birth of each child they may be legally forced to support.

Although these two ancient strategies will become much less successful, alternative strategies, uncommon today, will become more successful. For example, women will become much freer to have children by several men. Not only might they gain genetically, they might also gain from obtaining long-term financial support from a number of men (Scene 18). In the process of having children by several men, women should still seek to promote sperm warfare so that their male descendants will have competitive ejaculates (Scene 21). But no longer need women risk dependence on meagre support from the state as a result of their strategy. They will, of course, find it more difficult, if not impossible, to hide their actions from any partner, but then they will have less need to do so. In fact, they will have less need for a partner at all. Even if they have a partner, his desertion need not be as crippling as in the past. In their turn, men could exploit this potential earning power of women. No longer need they fear being tricked into raising another man's child – if they take on such a child it will be knowingly, having judged that they have more to gain than to lose (Scene 15). Often, they may even calculate that the extra income their partner could generate through having other men's children could actually help them to raise their own children more successfully.

Although such behaviour by men and women could make long-term relationships less common – and will certainly change their nature – such relationships will probably still happen. Men will still try to avoid sperm warfare by guarding a mate (Scene 9) and women will still try to recruit a partner to help with all the other, non-financial, aspects of raising their children (Scene 18). But the financial arrangements for child-raising and the ability to identify paternity could well shorten such relationships, removing as they do two of their main functions. The reproductive life of both men and women will centre on a succession of relationships, each lasting just long enough to produce one or two children.

As far as men are concerned, the balance of reproductive success is likely to swing even more strongly in favour of those of wealth and status than at present (Scene 18) – subconsciously, this may even be why child support legislation appeals to the law-makers! Only wealthy men will be able to afford to have children with a number of women, and hence only such men will be targeted by many women. Poorer men, when they do get a chance to inseminate extra women, will be under pressure to make themselves even more untraceable than at present.

Some things, of course, will never change. Nothing – short of castration, brain surgery or hormone implants – can remove a person's subconscious urge to have as many grandchildren as they can. So, nothing will remove a man's subconscious urge to have as many children with as many women as his genes and circumstances will allow. Similarly, nothing will prevent a woman from subconsciously trying to collect the best genes and recruit the best support for her children that *her* genes and circumstances will allow.

The coercive 'one-family, one-child' legislation in China has neatly highlighted how basic reproductive strategies adapt to social change. The legislation has successfully reduced the mean number of children (to 1.6 per woman). But in so doing it has also changed the sex ratio (through selective abortion and infanticide – Scene 16) to 1.6 boys for every girl. Why? Intrinsically, the coercion impinges more on a woman's potential reproductive success than a man's. A successful man (such as those who conceived and imposed the legislation?) can still have many children if he can surreptitiously inseminate many women and win sperm warfare. In contrast, the only way a woman can generate more grandchildren than her contemporaries is by having such a successful son. A man also, of course, benefits reproductively from fathering a successful son. Whatever people's conscious reasons for wanting sons so badly in modern China that they are prepared to kill their daughters, their biological response is precisely what would be expected if they were trying to enhance their reproductive success.

All of the sexual strategies described in this book will adapt to any new environment. Whatever the effect on society of future scientific and social developments, sperm warfare and its associated behaviour seem destined to be an ever-present feature. As such, it will continue to be the major force in the shaping of human sexuality over the generations to come.